D0387067

WITHDRAWN

WITHDRAWN

TO BREATHE FREE

TO BREATHE FREE

EASTERN EUROPE'S ENVIRONMENTAL CRISIS

EDITED BY JOAN DEBARDELEBEN

The Woodrow Wilson Center Press
Washington, D.C.

The Johns Hopkins University Press
Baltimore and London

Editorial Offices:
The Woodrow Wilson Center Press
370 L'Enfant Promenade, Suite 704
Washington, D.C. 20024

Order from:
The Johns Hopkins University Press
701 West 40th Street, Suite 275
Baltimore, Maryland 21211-2190
order department telephone 1-800-537-JHUP

© 1991 by the Woodrow Wilson International Center for Scholars

Printed in the United States of America
∞ Printed on acid-free paper

9 8 7 6 5 4 3 2 1

Library of Congress Cataloging-in-Publication Data

To breathe free : Eastern Europe's environmental crisis / edited by
Joan DeBardeleben.
        p.   cm.
   Includes bibliographical references and index.
     ISBN 0-943875-26-9 (alk. paper). — ISBN 0-943875-23-4 (pbk. :
   alk. paper) : $14.95
     1. Environmental protection—Europe, Eastern.
   2. Environmental policy—Europe, Eastern.   I. DeBardeleben,
   Joan.
   TD171.5.E86T6   1990
   363.7'00947—dc20   90–40073
                                                          CIP

WOODROW WILSON INTERNATIONAL CENTER FOR SCHOLARS
BOARD OF TRUSTEES

William J. Baroody, Jr., Chairman; Dwayne O. Andreas, Vice Chairman;
Robert McC. Adams; Lamar Alexander; J. Burchenal Ault;
James A. Baker III; James H. Billington; Henry E. Catto;
Lynne V. Cheney; Gertrude Himmelfarb; Eli Jacobs; John S. Reed;
William L. Saltonstall; Louis W. Sullivan; John H. Sununu;
Robert H. Tuttle; Don W. Wilson

The Center is the "living memorial" of the United States of America to the
nation's twenty-eighth president, Woodrow Wilson. The U.S. Congress
established The Woodrow Wilson Center in 1968 as an international
institute for advanced study, "symbolizing and strengthening the fruitful
relationship between the world of learning and the world of public affairs."
The Center opened in 1970 under its own presidentially appointed board of
directors.

In all its activities The Woodrow Wilson Center is a nonprofit, nonpartisan
organization, supported financially by annual appropriations from the U.S.
Congress, and by the contributions of foundations, corporations, and
individuals. Conclusions or opinions expressed in Center publications and
programs are those of the authors and speakers and do not necessarily reflect
the views of the Center staff, fellows, trustees, advisory groups, or any
individuals or organizations that provide financial support to the Center.

Woodrow Wilson International Center for Scholars
Smithsonian Institution Building
1000 Jefferson Drive, S.W.
Washington, D.C. 20560
(202) 357–2429

CONTENTS

TABLES

ACKNOWLEDGMENTS

East European Studies of the East and West European Program of The Woodrow Wilson Center would like to acknowledge assistance and funds, which made the June 1987 conference "Environmental Problems and Policies in Eastern Europe" and this volume possible, from a number of individuals and institutions. Financial support came from generous grants from the Rockefeller Brothers Fund to our program and from Federal Conference Funds to The Woodrow Wilson Center. U.S. Department of State Title VIII funding generally supports East European Studies and allowed several younger scholars to attend the conference. Gus Speth, president of the World Resources Institute, gave a fascinating talk about the greenhouse effect at the conference dinner. We would like to thank the World Watch Institute and the Conservation Foundation for their interest in the meeting.

Maya Latynski coordinated the effort of putting the book together, and Gertrud Stiefler provided invaluable editorial assistance. Finally, and most importantly, we would like to thank the authors—especially those from Eastern Europe—for their patience and their grants to us of liberties to abridge and edit their papers for publication in the days of an even greater technological gap.

PREFACE

This volume had its origins in a conference, "Environmental Problems and Policies in Eastern Europe," held at the Woodrow Wilson International Center for Scholars in Washington, D.C., June 15–16, 1987. Organized by the Center's East European Program and supported primarily by a grant from the Rockefeller Brothers' Fund, the meeting brought together environmental specialists from Poland, Czechoslovakia, Hungary, Bulgaria, Yugoslavia, Austria, and West Germany. They presented papers on the precise extent of air and water pollution in their countries, sometimes also outlining the official policies then in effect to address the grave problems that were apparent even from official East European data. We invited American specialists from the Environmental Protection Agency and a variety of private organizations to serve as commentators. Their comments helped to inform the revised versions of the European papers presented here.

Those revisions do not of course include recognition of the new political forces that swept Communist regimes from power throughout the region in 1989 and are now struggling to set up post-Communist governments and institutions. Nor need they do so. This volume does not attempt to be a current account of the state of environmental policy and official institutions in Eastern Europe. New institutions are only slowly taking shape. In the meantime, much of the old apparatus remains in place. The new leaders and parties have found it difficult to cover the economic cost or accept the political risk of imposing expensive environmental controls on the large industrial enterprises that are the principal polluters. In Poland and Hungary we see the real threat of a political backlash from workers facing unemployment when such enterprises lose even part of their state budget subsidy, let alone face new charges for pollution control or penalties for its absence. The separate environmental movement that played a prominent part in the overthrow of Communist power has not, moreover, survived as a powerful separate political party anywhere in Eastern Europe. Its chances appeared greatest in East Germany and Czechoslovakia but in neither place has the Green political organization expanded or even maintained its pre-1989 leverage.

What does remain the same throughout the region, however, is the a predictably worsening set of environmental problems and widespread public anxiety about the damage being done. The papers in the present volume record in unprecedented detail an amount and impact of air and water pollution that have not been reduced significantly in any East European country since 1987. The search for some feasible way to arrest this damage remains as compelling and yet, for different reasons, illusive as it was when the Communists were in power. Their overriding commitment to energy-intensive growth and their lack of public accountability are gone, even where some of them still occupy official positions. The legacy of environmental damage they left behind continues to hang over Eastern Europe. This volume serves to spell out that long-accumulating legacy and may thus contribute to its undoing.

John R. Lampe
Director of East European Studies
The Woodrow Wilson Center

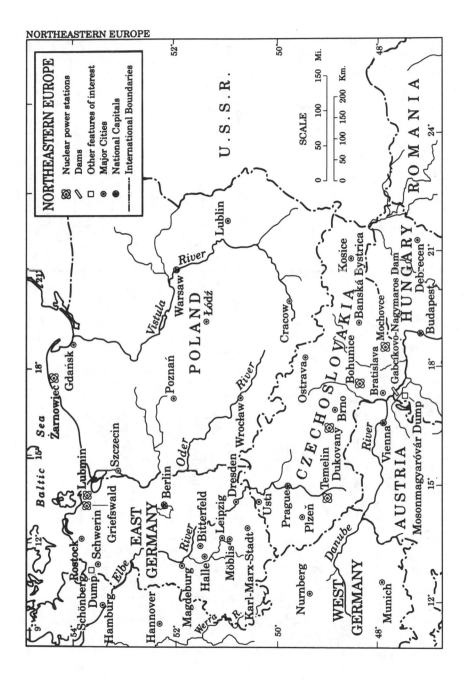

NORTHEASTERN EUROPE

- ⊗ Nuclear power stations
- ⟋ Dams
- ☐ Other features of interest
- ⊙ Major Cities
- ● National Capitals
- ·—·· International Boundaries

SCALE

Mi.
0   50   100   150

Km.
0   50   100   150   200

U.S.S.R.

ROMANIA

Baltic Sea

Żarnowiec

Gdańsk

Lubmin

Griefswald

Szczecin

Rostock

Schönberg Dump

Schwerin

Hamburg

Hannover

EAST GERMANY

Berlin

Magdeburg

Bitterfeld

Halle

Leipzig

Möblis

Karl-Marx-Stadt

Elbe River

Werra R.

Poznań

POLAND

Warsaw

Łódź

Lublin

Vistula River

Oder River

Wrocław

Dresden

Usti

Prague

Plzeň

Temelin

Dukovany

CZECHOSLOVAKIA

Ostrava

Cracow

Brno

Bohunice

Banská Bystrica

Košice

Mochovce

Bratislava

Gabcíkovo-Nagymaros Dam

Mosonmagyaróvár Dump

Vienna

AUSTRIA

HUNGARY

Budapest

Debrecen

Danube River

Nurnberg

Munich

WEST GERMANY

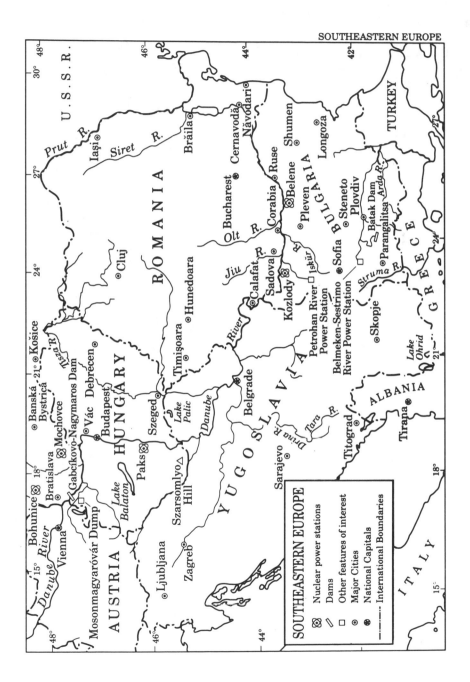

TO BREATHE FREE

INTRODUCTION

Joan DeBardeleben

It is by now a cliché in both East and West that the ecological crisis (environmental deterioration and resource depletion) is among the most pressing "global problems" facing humankind as a whole. Although environmental protection measures at local and national levels are necessary to maintain the quality of life in particular regions, they will be inadequate and often ineffective if not accompanied by a broader "global" strategy. For pollution and environmental damage respect no political boundaries; effects are often distant from sources of damage; and the larger political-economic setting has an important influence on the ability and willingness of individual countries or localities to commit scarce resources to environmental protection. Scientists, politicians, and environmental activists repeatedly call for international cooperation to solve increasingly urgent ecological problems such as acid rain, ozone depletion, the greenhouse effect, extinction of animal and plant species, and contamination of the food chain and water supply with pesticides and other pollutants. One should not forget, however, that successful international cooperation depends on the ability and willingness of individual countries to make commitments and devote resources to their fulfillment. Therefore it behooves us to gain a firmer and deeper understanding of the dilemmas confronted by our would-be partners in this global effort.

This volume is the product of a genuine international effort to grasp the domestic and international factors underlying the dramatic environmental deterioration that has been occurring in the countries of Eastern Europe. On 15 and 16 June 1987, scholars from Eastern Europe, Western Europe, the United States, and Canada met at the Woodrow Wilson International Center for Scholars in Washington, D.C., for a path-breaking discussion of the subject. Experts from each of the East European countries except Romania and the German Democratic Republic (GDR) were present. The organizers of the conference, The Woodrow Wilson Center's East European Program, attempted to fill these gaps by inviting Western scholars who are expert in the environmental situation in these two countries. The conference resulted in the publication of this volume. The participants sought not to produce manifestos or policy programs, but to improve understanding in both East and West of the dramatic challenge facing the East European societies. The conference offered an opportunity to reflect on questions such as these: What is the actual scope and nature of environmental problems facing these societies? How do political and economic structures inhibit or facilitate effective environmental protection? How do international and domestic forces interact to influence national policy-making? The discussion was open and honest, even when disagreements arose and approaches differed.

Following the conference all participants were encouraged to revise and update their contributions on the basis of subsequent developments and comments made at the meeting. East European participants facilitated the publication of the volume by allowing the editor and her assistants to make stylistic and organizational changes in their chapters, in order to make their content clearer to the English-speaking reader. In most cases the East European authors entrusted this responsibility to us without requiring final approval of the revised version. We trust that in the process we have not unwittingly introduced any stylistic revisions that alter the intent of the authors.

Since the authors completed the revisions of their chapters, dramatic changes have continued to unfold in the political relationships and regimes in several East European countries. Communist parties have been removed from power almost everywhere in Eastern Europe, and competitive elections have catapulted new political forces into power. The wall dividing East and West Germany has come down, and a reunified Germany must now confront the problems and issues that previously were the separate concerns of the FRG and GDR. Meanwhile, beneath these dramatic political occurrences profound social transformations are under way in all of the East European societies. While these changes have not yet had any marked effect on the environmental conditions in these countries, they have altered the political environment in which the debate over ecology will occur. Political forces supporting ecological platforms have become more vocal and visible, and many new public organizations have been formed at the grass-roots level.

Like other volumes on Eastern Europe in press at the time of these dramatic changes, ours, in some regards, is behind events even before its publication. Most of the chapters with a technical focus, however, are little affected by circumstances; and the chapters with a more political orienation still provide essential background and set the context for today's situation. Throughout this introduction, I will point out important trends that were emerging in the ecological area as Eastern Europe began a political and economic renovation.

THE POLITICAL-ECONOMIC CONTEXT

Although environmental problems facing Western and Soviet bloc countries are, in a technical sense, similar, their political and economic causes may be quite different; likewise, responses to the problems have varied. To some extent environmental deterioration seems to be an inevitable by-product of intensive industrialization and urbanization. Yet, as Chapter 1 by Barbara Jancar-Webster and Chapter 2 by John M. Kramer suggest, the character of the environmental dilemma is also affected by the specific path and strategy of development pursued.

The forced and rapid industrialization of the Stalinist variety imposed in Eastern Europe after World War II placed even greater stress on the environment there than in the USSR. The East European countries are generally much less richly endowed with mineral resources, open space, water, and other natural amenities than is the Soviet Union. Population density is generally higher. The expansive, "big is beautiful" mentality, in some ways appropriate to the burst of industrial development in the USSR in the 1930s, brought with it wasteful use of natural resources; an emphasis on quantity, not quality; and disregard for the long-term "external" effects of production decisions. Especially in the less developed areas—in Romania, Poland, Bulgaria, and Yugoslavia—urbanization has been rapid and poorly planned. The pressures on the urban environment caused by rapid industrial development have been only minimally ameliorated by effective industrial siting and timely pollution abatement measures. A cumbersome and centralized economic planning system encouraged waste of natural resources and highly inefficient use of energy in the productive processes, resulting in increased emissions of pollutants. The soil has been tilled intensively, overfertilized, and overchemicalized in an often disappointing effort to match crop yields with Western levels in order to satisfy rising consumer expectations. In the latter years of communist rule attempts to reduce foreign debt led political leaders to mandate increased reliance on low-quality domestic energy resources (mainly brown coal or lignite); the result was deteriorating air quality, acid rain, and damaged forests. And at the same time, citizens and politicians alike aspired to achieve Western levels of consumption, portending a shift to a consumer culture more akin to our

Western throwaway society. These and other pressures have combined to produce what many East European scholars recognize as an impending ecological crisis.

Overall, the East European societies lag behind their counterparts in Western Europe and North America in terms of level of industrial production, standard of living, individual consumption, and other measures of economic well-being. In some ways there are ecological advantages to this "backwardness." For example, there are fewer automobiles per capita, personal consumption is lower, and thus domestic waste is also smaller. At the same time, however, the rapid pace of development experienced in Eastern Europe in the last three decades has provided these societies with less time and fewer resources to introduce adaptive measures to lessen unanticipated environmental stress. Urban planning has often ignored environmental considerations, green areas have sometimes been destroyed before their value was appreciated, and choices in development were introduced rapidly and broadly before their long-term ecological consequences could be foreseen. Furthermore, as these societies seek to emulate Western levels of production, there are few free resources left to devote to the elusive good that is environmental quality.

Evolution of the Environmental Movement

Since the mid-1960s an ecological awakening has occurred in the USSR and Eastern Europe. The "big is beautiful" mentality has come under severe attack in the USSR and in several of the East European countries. The notion that nature is inexhaustible has succumbed to the vulnerability of even Siberia's great expanses. In Chapter 4, Philip Lowe provides a comparative counterpoint for examining this awakening in Eastern Europe by identifying important sources for the ascendancy of environmental politics in Western Europe. He attributes key importance to the shift in the pattern of West European industrial and economic development since World War II, involving a larger service industry and greater emphasis on leisure activities as a basis of production and enjoyment. In this context, a clean and healthy environment has become a scarce but highly valued social commodity.

Broad public awareness of environmental quality as a scarce and valued good came later to Eastern Europe. Other problems undoubtedly seemed more immediate to the average citizen: the meager supply of high-quality consumer goods, housing shortages, and the infrequent availability of fresh fruits, vegetables, and good meat. The service sector and leisure industries have remained underdeveloped until this day. The shift in the production structure that Lowe identifies in Western Europe is beginning to occur only now in Eastern Europe, particularly as economic reforms in countries like Hungary and Poland are allowing the private sector to cater to the desires of the population for scarce goods and services. At the same time, the East European public has gradually gained increasing exposure to environmental politics in Western Europe, particularly with the important and visible position taken on by the "Greens" and by antinuclear groups in several neighboring countries, most notably West Germany. Not only East Germans but also many Hungarians, Czechs, and Slovaks were able to receive and understand West German or Austrian radio and television broadcasts even before the recent openings in the border; thus they were able to broaden their knowledge of ecological tendencies and movements in the West. Since the dramatic developments in Eastern Europe in 1989, this type of exposure to Western ideas and experience has increased dramatically. Here, as in some other spheres, Eastern Europe has served as a harbinger of developments in the Soviet Union by providing a conduit for information from the West.

As Barbara Jancar-Webster (Chapter 1), Miklós Persányi (Chapter 11), and I (Chapter 9) document, a real ecological movement of sorts developed in the 1980s in several East European countries, most notably Poland, East Germany, Hungary, and Yugoslavia. In the USSR as well, increasing popular activity has emerged in the last five years. The April 1986 accident at the Chernobyl nuclear power plant clearly has served as an important catalyst in this process. The invisible cloud of radiation that spread over Eastern Europe from the Ukraine made the cost and fallibility of advanced technology evident to large sectors of the East European public. Furthermore, the East European countries appeared to be victims of Soviet carelessness. Ecological activism could subtly join nationalist sentiments. Meanwhile, the long-standing commitment in Eastern Europe and the USSR to nuclear power as a "clean" energy resource, while not officially repudiated, has been placed in question. On safety grounds, the new German government had decided to close the Soviet-type reactors previously in operation in the GDR, and Poland has also reversed the former regime's commitment to development of nuclear power. In other countries, such as Czechoslovakia, reliance on nuclear power will probably continue, but with improved safety standards and equipment. Several of the authors in this volume, including John M. Kramer (Chapter 2), Winfried Lang (Chapter 5), Helmut Schreiber (Chapter 6), Edward Mainland (Chapter 13), Mihailo Crnobrnja (Chapter 14), and me, examine dimensions of the impact of Chernobyl on various aspects of East European environmental politics in the 1980s.

In Chapter 1, an insightful discussion of groups affecting environmental politics in Eastern Europe in the 1980s, Jancar-Webster attributes special importance to experts and intellectuals as key elements of the ecological lobby. In fact, in comparison to Western countries, scientists played an even more pivotal role in the early stages of ecological awareness. For it was they who had access to the required information and knowledge that enabled them to recognize the impending dangers in the early-to-mid-1960s. Joined by journalists, writers, philosophers, and social scientists, natural scientists were gradually able to inject their warnings into more widely disseminated media. In many cases, they also had better access to policymakers than the average citizen; the regimes turned to them for advice in formulating economic plans and environmental policy. Experts are still an important source of information for struggling unofficial ecological groups that are now cropping up in the USSR and Eastern Europe. They can provide insight and understanding, as well as data that are not otherwise available to the public at large. Although scientists are also an important component of the environmental lobby in the West, the movement there gained a broader and more vocal public following much earlier than in the East bloc.

Another factor spurring the environmental awakening in Eastern Europe is the sheer intensity of the problem. Several of the East European authors whose "country studies" appear in Part III document the extent of environmental deterioration. Some is "invisible" to the average citizen, but much damage leaves marks recognizable even to the poorly informed layperson. For example, acid rain is having an increasingly visible effect on European forests, especially in East Germany, Czechoslovakia, and Poland. Trees are literally dying before one's eyes. Likewise, the reemphasis on brown coal as a major source of energy, which occurred in the late 1970s, meant the expansion of strip-mining, another very visible eyesore. More beaches are closed to swimming, fish kills are more frequent, and more green areas are taken over by "development." With the increase in personal automobile ownership, urban air is even more obviously unfit for human respiration. East European citizens, through various sources, also hear about the global side of the dilemma—depletion of the ozone layer, the greenhouse effect, and the ubiquitous presence of harmful chemicals in the food chain and environment. Lowe's thesis—that general awareness combined with personal exposure to environmental degradation increases the likelihood of environmental activism—is becoming a reality in many areas in Eastern Europe, especially in the more developed countries (the former GDR, Hungary, Czechoslovakia, and Poland), where Western contacts are the most extensive.

Until recently, it made sense to distinguish between "official environmentalism" (positions on the environment endorsed by official experts and authorities) and "unofficial environmentalism" (positions put forth by unofficial organizations or by individuals who have no official status). As Charles Ziegler notes in Chapter 3, on the official level a long-standing concern with careful husbanding of resources has supported efforts to improve efficiency of production. In countries such as the GDR this concern spilled over into a commitment to an assertive recycling policy. While recognizing that nature imposes limits, the dominant ideology always reflected a firm faith in technology and science as tools to address the problems. Official environmentalism also consistently affirmed the primacy of economic growth as a policy goal. Some established economists tried, however, to demonstrate that commitment of resources to environmental protection need not draw resources away from growth-producing sectors; on the contrary, if long-term growth patterns are to be sustained, state policy must also assure that the regenerative capacity of the natural environment is retained. By taking this type of approach, economists attempted to persuade leaders that an assertive environmental policy may further other policy priorities. In Chapter 9, I try to illustrate the development of this type of approach by focusing on the example of the former GDR.

The communist governments also sponsored the formation of "official" public organizations to provide support for state policy and to channel popular concern into acceptable activities. For example, in the face of rising environmental activism in the GDR, the Society for Nature and the Environment was founded in 1980; by 1989 it had some sixty thousand members.[1] Likewise, in October of 1989 as environmental awareness was growing in Bulgaria, an Ecology national youth club was formed under the Central Committee of the Komsomol.[2] Jancar-Webster discusses organizations of this type more fully in Chapter 1.

Creation of these groups, however, failed to head off the emergence of a significant "unofficial" environmental movement at the grassroots level over the course of the 1980s in virtually every country in Eastern Europe. The unofficial groups gained considerable momentum in the first half of the 1980s in the GDR, Hungary, and Poland; developments were slower in Czechoslovakia, although some activity emerged within the Charter 77 organization. Not until 1987–88 did

Bulgarian activists become visible. Evidence of grass-roots environmental activism in Romania began to emerge only in late 1989.

In most countries the momentum has increased even more rapidly in the late 1980s. Unofficial environmental groups have been a factor in the popular self-assertion that has brought about the decline of Communist party dominance in the East European countries; grassroots environmentalism will also be a major beneficiary of these developments as possibilities for independent action expand.

The unofficial groups of the 1980s were in some cases directly, in other cases implicitly, linked to other political tendencies. A prime example is the association of ecological concerns with nationalist sentiments. For instance, in the USSR, defense of the integrity of the Siberian motherland motivated opposition to the vast river diversion projects proposed to move water from Siberia to Central Asia. Implicitly, this involved an assertion of Russian nationalism vis-à-vis Central Asian interests. Armenian protests over patterns of control in the Nagorno-Karabakh region and nationalist pressures in the Baltic states have also contained a strong element of ecological nationalism, that is, a protest against the violation of a pristine national environment by outside influences. Likewise in Eastern Europe, as was noted, reaction to the Chernobyl accident included an element of strong resentment against "foreign" intrusion. This nationalism underlying some strains of environmentalism has afforded them important popular resonance and has also likely helped reinforce the commitment of East European activists to forging an indigenous economic strategy. In this sense, the fledgling ecology groups that existed before the recent revolutionary eruption have helped to lay the groundwork for renewal in Eastern Europe.

Likewise, the "unofficial" environmental groups were in some cases (as with Charter 77 in Czechoslovakia and GDR ecologists in the Evangelical church) also closely linked to other "critical" or dissident strains of thought in Eastern Europe. Independent ecologists often posed challenges to the Stalinist economic and political model, with its priority on heavy industrial (and thus polluting) production and its grandiose industrial gigantism. In this way, these groups also helped set the stage for the radical rejection of the modified Stalinist model that prevailed until recently in Eastern Europe.

With the increasing delegitimation and, in some cases, replacement of Communist party rule in Eastern Europe, what used to be "unofficial groups" have become legitimate participants in the political arena. They are now able to play an active and open role in shaping the political agenda and in some cases have even taken on the character of aspiring or self-proclaimed opposition political parties. For example, in the months immediately following the resignation of the old-guard leadership of the SED in the GDR, a new Green party was formed.[3] In 1990 it joined forces with the West German Greens and other citizens' movements in the East to form "Bundnis 90" to compete in the all-German elections on December 2, 1990. The coalition won only 1.2 percent of the all-German vote, but claimed 5.9 percent in the former GDR, entitling the group to seventeen deputies (all from the East) in the new Bundestag. A Polish Green party (until December 1988, the Polish Ecological Party) has been in existence since Septmeber 1988, when it was founded in Cracow.[4] Attempts to establish a Green party (Party of the Green Masses) late in 1988 in Bulgaria were less successful, as they were subject to almost immediate government repression.[5] In Hungary at least two of the independent groups in the Opposition Round Table demonstrated a significant commitment to environmental matters and helped to raise popular consciousness of the issue (the Bajcsy-Zsilinszky Society and the Hungarian Democratic Forum).[6] Even in Romania, a type of Green party (the Romanian Ecological Movement) was formed in late 1989. While not yet major actors at this stage, the very existence of the Green parties exemplifies the shift from the marginalized unofficial status of the ecology movement to its role as an independent political actor. Alongside these fledgling parties are numerous other grass-roots organizations, some of which are more broadly based in their respective countries and could thus in the future form the basis for Green parties. Some of the newer groups will be discussed next. In the context of this burgeoning public activism, the role of environmental considerations in establishing public policy relating to issues such as future energy policy, strategies for agricultural and industrial renewal, and general programs for economic reform, will now be subject to a much more thoroughgoing and informed public debate.

Many of the new grass-roots movements are at the local level and thus have gone virtually unknown in the West. Others have gained more prominence and deserve mention here. In January 1988 a Green Ark Network was formed in the GDR, established to help coordinate activities among the numerous ecological groups which had sprung up under the umbrella of the GDR Evangelical (Lutheran) church. (Radio Free Europe Research estimates that in June 1989 about five hundred base groups, dealing with environmental issues in local communities, existed in the GDR.)[7] In January 1989, GDR officials met with representatives of the Green Ark Network to discuss problems associated with disposal of hazardous waste from the Federal Republic of Germany (FRG) at the Schöneiche site.[8] This event not only suggested the growing political weight of the GDR ecological movement on the eve of reunification but also exemplified a more open approach to discussion of ecological issues demonstrated by the old GDR leadership in the last year of its rule.[9]

Developments in Bulgaria were perhaps even more rapid than elsewhere. As noted in Chapter 1, first protests about chlorine pollution (originating across the border in Romania) in the city of Ruse apparently began in 1987. In March 1988 an Independent Committee for the Protection of the Environment (the Ruse Committee) was formed, subsequently subject to considerable official harassment.[10] In early 1989 an offshoot of the Ruse Committee took on a broader scope with the founding of Eco-Glasnost.[11] As its name suggests, this national organization is publicly committed to expanding public information and stimulating public awareness of ecological problems in Bulgaria. Among its specific concerns have been the proposed water management projects in the Rila Mountains, dumping of radioactive waste in Bulgaria, and pollution of the Black Sea. Eco-Glasnost has begun to forge a whole series of international connections to buttress its position at home. These include contacts with other dissident groups within Bulgaria (as well as overlapping membership in these groups); meetings with environmental activists from the former GDR, Poland, and the USSR; and discussion with the international Greenpeace organization. From 16 October to 3 November 1989 Bulgaria acted as host to the International Ecoforum, a meeting on the environment held as part of the Conference for Security and Cooperation in Europe (CSCE) process. This event offered Eco-Glasnost the opportunity to gain international recognition through its very visible public profile as a sponsor of petition campaigns and through its daily public presence at the Kristal Park in central Sofia. During the first week of the Ecoforum, Bulgarian authorities were reluctant to take overt action against Eco-Glasnost, and its activities proceeded relatively uninhibited. A clampdown did follow on 24 October, however.[12] With the fall of the Zhivkov government under popular pressure in December 1989, Eco-Glasnost has become a visible actor in the national political process unfolding in Bulgaria.

In other East European countries there have also been new initiatives. In Czechoslovakia in early September 1989, the Northern Moravian Ecological Club and a Green Club in northern Bohemia were formed.[13] Protests against the joint Czechoslovakian-Hungarian Gabčikovo-Nagymáros Dam project on the Danube have taken place in

Hungary, with some minimal resonance in Czechoslovakia (these will be discussed below). In Romania, some incipient environmental activism also has occurred.[14]

As open political competition is increasingly becoming the norm in Eastern Europe, the distinction between official and unofficial environmental groups has broken down, and the new initiatives should more appropriately be termed "independent" rather than "unofficial." In the future, in most East European countries, political leaders will most likely have to accept the legitimacy of a broad range of environmental groups and respond to them in a political context involving numerous and often conflicting priorities and popular demands.

Already several of the new regimes have, by action or words, committed themselves to a more open disclosure of environmental information or to a more assertive environmental effort. For example, officials of the GDR Office of Atomic Safety and Radiation Protection released information about an accident at the Lubmin nuclear power plant in the north of the GDR in 1976.[15] This information likely intensified anxiety in the GDR and elsewhere in Eastern Europe about plans to expand nuclear power production in the coming decade, particularly because the reactor involved in the accident was a VVER—water moderated reactor—the type operative throughout Eastern Europe, which had been thought safer than the RBMK—graphite moderated reactor—the kind used at Chernobyl, which has never been exported to Eastern Europe.[16] Meanwhile, in Hungary by early May 1989 public pressure led the government to announce its withdrawal from the controversial Gabčikovo-Nagymáros Dam project (with cessation of work on the Nagymaros Dam located in Hungary).[17] The decision was confirmed by the Hungarian parliament on 31 October 1989.[18] The Hungarian pull-out from the project led to opposition between the governments of Czechoslovakia and Hungary, since the former sees its portion of the project (including the planned start-up of the Gabčikovo hydroelectric project) threatened by the Hungarian decision and is demanding compensation.[19] The new Czechoslovak government may well be more sympathetic than the old to the democratic process that led to Hungary's withdrawal, and more public debate may ensue in Czechoslovakia as well. The project at Gabčikovo is, however, very close to completion, and therefore a reversal is unlikely. Furthermore, in Czechoslovakia no strong public movement similar to that which pressed the Hungarian government to reverse its position has ever existed.

These types of developments are only one small indication of the increased saliency that environmental issues are likely to take on in Eastern Europe, now that expanded information and largely unrestricted opportunities for organization exist.

Economic Reform and the Environment

Many Soviet bloc and Western economists have emphasized the importance of market mechanisms in promoting effective environmental policy in Eastern Europe. In Chapter 2, John M. Kramer points out that the artificially low prices on oil exports from the USSR to Eastern Europe in the late 1970s and early 1980s reduced the incentive to conserve energy. World-level market prices would have made it economically rational to cut back on energy use and reduce waste, as occurred in the West in the same period. Along similar lines, both Western and some Soviet bloc economists have argued that fees for using natural resources within individual countries are necessary to encourage enterprises and other users to use natural resources more carefully. Even under communist rule, some incipient efforts in this direction were made in several East European countries and in the USSR; the effectiveness of those efforts was, however, hampered by the pricing system as a whole and by other limits on enterprise autonomy.

Now key questions for environmentalists in Eastern Europe involve the impact that market reforms will have on environmental quality and whether other pressing social and economic problems will divert attention and resources away from ecological concerns. Soviet bloc spokespersons have traditionally claimed that a centrally planned economy should be more capable than capitalist economies of addressing complex problems such as environmental deterioration; these claims certainly were not borne out in practice. On the other hand, market mechanisms in Western countries have not provided effective safeguards against environmental deterioration either. In market systems, self-interested decision making most often has *not* added up to environmentally responsible decision making. While market signals and pressures may have encouraged individual firms to reduce the waste of resources, they have not discouraged the same firms from sloughing off the costs of pollution onto society. Furthermore, the imperative to growth that is intrinsic to market competition has a tendency to encourage conspicuous consumption and production of throwaway commodities, forces that are not at all inherent in a centrally planned system. On the other hand, market pressures do provide incentives for efficient utilization of energy and natural resources, which in turn reduces polluting emissions. Inefficient use of energy and raw materials has proven to be a costly weakness of the traditional Soviet economic mechanism, and experience has also shown that costs of production from pollution were often externalized in Soviet bloc countries. Neither economic system has found an easy way to build safeguards against such social costs into the economic structure. In the West, political pressure from independent scientists and citizen groups and the resultant regulatory legislation—forces that have less to do with economic systems than with political decision-making structures—sometimes do provide mechanisms that were weak indeed in the former Soviet bloc countries.

The real issue may well be the *priority* of environmental protection as a policy goal, whether this be in centrally planned systems or in market-based systems. Therefore, I am less sanguine than some of the other authors about the likely contribution of market-type reforms in furthering environmental protection in Eastern Europe. More important will likely be the possibilities offered by more open public dialogue and public activism under conditions of genuine political competition. Also important will be the particular commitment of the governing leadership to the issue. Jancar-Webster addresses the latter point in Chapter 1 in regard to the Gorbachev generation.

THE INTERNATIONAL FACTOR

International cooperation is a particularly important dimension of environmental protection in Eastern Europe. The several fairly small countries in this region share river streams, coastlines on water bodies, atmospheric conditions, border regions, and also, of course, pollution of these resources. Conditions in Eastern Europe are dependent on pollutants emanating from Western Europe and, because of prevailing winds, less frequently from the USSR. As Charles Ziegler shows in Chapter 3, a whole series of cooperative efforts and agreements between the USSR and the East European countries were undertaken within the context of the Council for Mutual Economic Assistance (CMEA) in the 1980s. With the dismantling of the CMEA in early 1991, cooperative efforts now are likely to take on a more all-European character or center around limited bio-geographical regions which cut across the former division between Eastern and Western Europe. Internally, individual countries are implementing various types of economic reforms which may have an indirect or direct impact on the quality of the environment[20] and thus also allow more diverse forms of international cooperation.

The Soviet proposal to account intrabloc trade in hard currency at real market prices was put in effect at the beginning of 1991, despite earlier opposition from several East European countries.[21] This has produced a general fall in the trade turnover between many East European countries and the USSR, as both sides have sought more promising markets and suppliers. Most painful for the East European countries is, however, the necessity to pay for Soviet oil and natural gas with scarce hard currency. Previously the USSR exported substantial amounts of oil and natural gas to Eastern Europe in exchange for ruble-valued East European goods. The transfer to hard currency accounting has placed short-term stresses on the economies of the region, which may make ecological compromises likely. Some East European countries may have to cut back imports of oil and natural gas even further, and this may necessitate continued reliance on heavily polluting indigenous coal or on nuclear power, which also raises environmental concerns.

As public policy becomes increasingly responsive to popular concerns and as Soviet influence declines, one may also anticipate conflicts between East European countries over environmental questions. In the past four decades, potential conflicts have in this policy area, like others, been muted by Soviet dominance in the region. The conflict between Hungary and Czechoslovakia over the Gabčikovo-Nagymáros Dam project may be indicative of the type of disagreements that could ensue. It is important that forms of conflict resolution, such as mediation and, possibly, arbitration, be activated.

Winfried Lang, a leading figure in the field of international environmental law, elaborates the considerable obstacles to binding agreements between nations in this field in Chapter 5. The predominance in this area of unenforceable "soft" law, as opposed to binding international agreements, makes it difficult to achieve marked gains. The nations of Eastern Europe, like other countries, have often been reluctant to commit themselves to joint efforts that may further strain already scarce resources. Lang suggests, however, that regional agreements addressing limited and specific issues have offered the most effective approach to international cooperation. Examples include efforts to protect the Baltic Sea and the commitment by several European nations to reduce sulfur dioxide emissions by 1993 to a level 30 percent lower than 1980 levels. Imrich Daubner of Czechoslovakia provides in Chapter 7 yet another concrete example of regional cooperation, this time involving protection of one of Eastern Europe's most important water resources, the Danube.

Helmut Schreiber explores a unique case of international cooperation, that involving the two Germanies, in Chapter 6. The German-German case was affected by the special relationship between the two countries. The government of the FRG had a unique interest in assisting its neighbor. West Germany and West Berlin "imported" significant amounts of both air and water pollution from East Germany and would have been greatly affected by any environmental "mishaps" near the common border. Furthermore, West Germany's *Ostpolitik* had, since the early 1970s, recognized the importance of forging close ties between the two countries on multiple levels (economy, culture, and personal relations between citizens) in order to facilitate the maintenance of a common German identity, ease the personal difficulties posed by split families and separated friends, and encourage change within the GDR. Financial assistance from the FRG for environmental protection projects in the GDR has provided East Germany with possibilities unavailable to the other countries of Eastern Europe.[22] This type of assistance paved the way for future environmental initiatives within a reunified Germany, as the new German government seeks to apply common standards to industrial producers in both parts of the country.

As several of our authors point out, the ability of individual countries to accept and adhere to international commitments depends ultimately on domestic factors: the priority of environmental protection at home, the availability of financial resources and trained personnel to address the problems, the existence of a public constituency to oversee efforts, and the presence of effective political and economic structures to realize commitments.

COUNTRY STUDIES

In Part III of this volume, entitled "Country Studies," the reader will find chapters dealing with each of the East European countries except Czechoslovakia. The two Western authors represented in this section, Edward Mainland (Chapter 13) and I (Chapter 9), provide overviews of environmental problems and policies in Romania and the former GDR respectively. (Unfortunately, invited guests from these two countries did not attend the conference.) The GDR and Romania represent opposite ends of the developmental spectrum in Eastern Europe: the GDR was, by most criteria, the most economically advanced country in the bloc, whereas Romania, along with Bulgaria, was among the least developed. Although specific comparative data are not available, these two chapters make clear that both countries had—and will continue to have—serious environmental problems. Official secrecy, restrictions on public debate and popular involvement, priority for traditionally conceived economic growth, and foreign debts were features that have traditionally hindered the implementation of an effective environmental policy in both countries under communist rule. In the GDR, the burning of low-quality domestic brown coal was the outstanding problem. Edward Mainland provides particular insight on the degeneration of the Danube delta in Romania. Mainland poses the provocative question of how one should really understand concepts such as "development," "growth," and "progress." For all its "development," was the GDR really any more environmentally progressive than Romania? In terms of legal enactments and regulations, it almost surely was; the GDR had also made notable progress in terms of recycling resources. Environmental awareness may also have been greater in the GDR than in a less developed country such as Romania. The level of environmental awareness may, however, have been even more difficult to assess and less visible in Romania than in the GDR because this awareness was restricted to local environmental conditions, with little recognition of larger implications or less visible forms of pollution (a pattern Lowe finds in Chapter 4 to be characteristic of the less developed countries in the southern part of Western Europe). On the other hand, writers, scientists, literary figures, philosophers, and young people working in the GDR in the Evangelical church since the mid-1970s were publicly presenting images of a more genuinely ecological society. The questions that Mainland poses in his chapter—questions about how to conceive "progress"—are equally relevant to the emerging environmental policies of the new Germany as they were to the former GDR. Indeed, all of Eastern Europe now stands at a crossroads; in constructing their new postcommunist societies and economies, these countries may seek to imitate Western conceptions of "progress" (including those aspects that are insensitive to ecological concerns) *or* they may try to learn from mistakes made in the West and leapfrog over the path to define a more ecologically sound understanding of needs, consumption, and production.

The country studies provided by East European authors generally have a more technical focus and deal with a specific dimension of environmental protection in each country. The exception is Chapter

11, titled "Social Support for Environmental Protection in Hungary," by Miklós Persányi. This piece provides an insider's perspective on popular involvement in Hungary before the fall of the communist regime. It demonstrates that environmental awareness was growing even then, as could be seen in the popular activism surrounding issues such as the protection of the Budapest hills and the Szarsomlyo hill in southern Hungary, reduction of noise pollution, contamination of supplies of drinking water, and importation and storage of hazardous wastes. Persányi notes the continued necessity of popular involvement, for serious environmental problems remain in Hungary. Also focusing on Hungary, Zoltán Király clarifies in Chapter 10 an environmental dilemma in agricultural policy that also faces other East European countries. Continued improvements in the Hungarian standard of living and diet are predicated in part on an improved food supply. But the traditional reliance on intensive use of pesticides and fertilizers is taking its toll on the environment. Alternative methods, relying more heavily on biological controls, are under study in Hungary but have not yet been widely applied. The willingness and technical capacity to shift to these types of methods are still not adequately developed either in Hungary or in the other East European countries. Meanwhile, nitrate levels in water and food are rising, pesticides in the food chain present an increasing threat to health, and the biological capacity of the soil is declining.

In Poland, information about the quality of the environment was, under communist rule, somewhat more available than in other East European countries (excluding Yugoslavia). Therefore Chapter 12 by our Polish contributor, Jerzy Kurbiel, provides a much clearer depiction of environmental quality in the 1980s than was generally available for the other countries of the region. During the period of activism of the Solidarity trade union in 1980–81, concrete data were published and a broader public became involved in unofficial environmental organizations, such as the Polish Ecological Club. Independent environmental activism was frowned upon by the Polish government from 1981 until the fall of the communist government in 1989. Nonetheless, even during that period information was more readily available than elsewhere in Eastern Europe. Beginning in mid-1989, even more information about the environmental situation was being released in Poland.[23] The quality of the environment has thus far benefited little, however. Jancar-Webster notes in Chapter 1 some intermittent successes, and further plant closings due to ecological considerations have been announced since her chapter was written.[24] However, the pervasive economic and foreign-debt crisis, as well as Poland's continued reliance on ever-poorer domestic coal for energy, have placed continuing stress on the environment. Jerzy Kurbiel provides a graphic depiction of the poor state of Poland's water resources. He also summarizes relevant legislation, describes institutional structures, and identifies major problems for water protection policy.

Water shortages plague several East European countries. In this context, Kurbiel (Chapter 12) and I (Chapter 9) discuss the importance of water recycling and purification in Poland and the former GDR respectively. Likewise, in Chapter 8 on Bulgaria, Georgi Gergov draws attention to the constraints posed by that country's drought-prone climate. Gergov provides a detailed description of available water resources and of major users of the limited water supply. Although detailed data on the quality of water in Bulgaria are not available, Gergov's wide-ranging discussion of the diverse problem areas suggests the scope of the challenge confronting Bulgaria's water management program, with its numerous water reservoirs, extensive irrigation system, and hydroelectric facilities.

In Chapter 14, Mihailo Crnobrnja examines the economic and environmental considerations underlying Yugoslavia's energy policy. Despite the various market and self-management reforms undertaken in Yugoslavia in the past four decades, Crnobrnja's chapter suggests, Yugoslavia still suffers from many of the same problems confronting the other East European economies. Market mechanisms have generally been distorted by governmental regulation, making the price system often unresponsive to signals of supply and demand. As in other East European countries, Yugoslavia's energy sector is characterized by high levels of energy consumption per unit of output, involving considerable energy waste. Recent reforms have sought to generate energy prices that will more accurately reflect opportunity cost and world price levels. But Crnobrnja expresses considerable doubt about the likely success of these reforms. Yugoslavia confronts another difficulty common to other East European countries, namely a burdensome level of foreign debt, which in turn constrains energy options.

One area where the Yugoslav situation does differ considerably from the other countries of Eastern Europe involves the extensive decentralization of decision making allowed by Yugoslavia's federal structure. Crnobrnja points out that those republics most richly endowed with energy resources are also the least developed economically, and thus they find themselves in a weak position to finance the exploitation of available resources. Cooperation between the republics has been inadequate to assure optimal use of energy resources on a national level. On a more hopeful note, Crnobrnja observes that in placing priority on the development of hydroelectric power, Yugoslav energy policy embraces the option that poses the fewest potential hazards for the environment.

CONCLUSION

Despite the diverse focuses of the contributions to this volume, one is struck by the similar problems confronting the various countries. This similarity is perhaps not surprising, given those countries' parallel paths of development following World War II and their close geographic proximity. Even before the events of 1989 we could, however, observe diverse models emerging to deal with the impending crisis. The GDR, with its impressive body of legislation and regulation, sought to limit environmental deterioration by applying largely administrative means of control within a traditional, centrally planned economy. Yugoslavia, on the opposite extreme, sought to apply some market principles, significant decentralization, and self-management to these problems, as to others. Finally, in Hungary application of limited market mechanisms within a socialist framework was the chosen path. None of these approaches proved markedly superior in improving environmental quality.

Under present conditions, the diversity of approaches will most likely expand even further. While decentralization of decision making and inclusion of market features in the economic structure seem to be common tendencies in the reform packages of the various countries, the mix of market and regulatory measures is likely to differ significantly from country to country, just as it does among West European countries. What we may hope to see is a real debate about the efficiency of market mechanisms as a response to environmental deterioration and about the priority that should be granted to environmental protection vis-à-vis other policy concerns. Environmental protection as a policy goal must be reconciled with the continuing drive for higher rates of economic growth and expanded consumption. Unless a critical evaluation of societal priorities takes place in earnest, efforts applied in all economic models will likely falter. For each type of economic structure has its weaknesses as well as its strengths in addressing environmental problems. The costs of any serious commitment to environmental protection must be accepted before any of the institutional and economic mechanisms can be expected to operate effectively.

From this perspective, the most encouraging sign for the environment in Eastern Europe is the broader public awareness of the problem and the growing willingness and ability of citizens to organize themselves to press for better solutions. As in the West, most likely the average citizen in Eastern Europe remains either unduly optimistic or resigned about the future of the environment. Only a continued dialogue on global, national, and regional levels will elicit the consciousness necessary to alter our priorities before the bell tolls for us all.

NOTES

1. Barbara Donovan, "New Resolve to Combat Ecological Decay in the GDR," *Radio Free Europe Research* (hereafter *RFE Research*), RAD Background Report/31 (German Democratic Republic), 14, no. 8, part I (21 February 1989), p. 1.
2. Vera Gavrilov, " 'New' Proposals on Environmental Protection Published," *RFE Research*, Bulgarian SR/10, 14, no. 49, part I (5 December 1989), p. 33.
3. For a statement on the founding of the GDR Green party, see "Gegen Ellbogenfreiheit, Verschwendung, Wegwerfmentalität," in *DDR Journal zur Novemberrevolution, August bis Dezember 1989, Vom Ausreisen bis zum Einreissen der Mauer* (Berlin: *Die Tageszeitung [taz]*, 1989), p. 85.
4. See Jiri Pehe, "An Annotated Survey of Independent Movements in Eastern Europe," *RFE Research*, RAD Background Report/100 (Eastern Europe), 14, no. 24, part I (14 June 1989), p. 24.
5. See Stephen Ashley, "An Attempt to Found a 'Green' Party," *RFE Research*, Bulgarian SR/1, 14, no. 5, part II (3 February 1989), pp. 15–17. See also Henry Spetter, "The Ecological Crisis in Bulgaria: No Solution in Sight," *Environmental Policy Review* 3, no. 1 (January 1989): 29–33.
6. See Zoltan D. Barany, "Hungary's Independent Political Groups and Parties," *RFE Research*, RAD Background Report/168, 14, no. 3 (12 September 1989), pp. 3–5.
7. Pehe, "An Annotated Survey," 10–11.
8. Donovan, "New Resolve," 4.
9. See, for example, the more-extensive-than-usual treatment of environmental protection in the 1989 economic plan. "Gesetz über den Volkswirtschaftsplan 1989 vom 14, Dezember 1988," *Neues Deutschland*, 16 December 1988. See also Donovan, "New Resolve."
10. Pehe, "Annotated Survey," 3.
11. Kjell Engelbrekt, "Bulgaria's Independent Environmental Movement: Eco-glasnost'," *RFE Research*, Bulgarian SR/1, 14, no. 32, part I (7 August 1989), pp. 15–18.
12. For an account of these events see Kjell Engelbrekt, "Activity among Dissident Groups during the First Week of European Ecoforum in Sofia," *RFE Research*, Bulgarian SR/10, 14, no. 49, part I (5 December 1989), pp. 37–43. On Eco-Glasnost's international and dissident contacts, see also Stephen Ashley, "Dissident Groups Forge Foreign Contacts," *RFE Research*, Bulgarian SR/10, 14, no. 49, part I (5 December 1989), pp. 11–13.
13. Jiri Pehe, "Independent Activity Continues, Grows despite Pressure from the Authorities," *RFE Research*, Czechoslovak SR/21, 14, no. 42, part IV (20 October 1989), pp. 28–29.
14. On Romania see "Underground Ecological Activity in Rumania," *Environmental Policy Review* 3, no. 1 (January 1989): 41–43.
15. *New York Times*, 23 January 1990.
16. For a discussion of the two types of reactors see Joan DeBardeleben, "Esoteric Policy Debate: Nuclear Safety Issues in the USSR and GDR," *British Journal of Political Science* 15 (1985): 227–53.

17. *New York Times*, 19 May 1989.
18. See Vladimir V. Kusin, "Gabcíkovo-Nagymaros: The Politics of a Project," *RFE Research*, RAD Background Report/206 (Eastern Europe), 14, no. 47, part I (24 November 1989), pp. 1–5.
19. See ibid.; and Peter Martin, "The Gabcíkovo-Nagymaros Dispute Intensifies," *RFE Research*, Czechoslovak SR/20, 14, no. 40, part IV (6 October 1989), pp. 19–23.
20. For an analysis of recent changes in environmental policy in the USSR see Joan DeBardeleben, "Economic Reform and Environmental Protection in the USSR," *Soviet Geography* 21 (April 1990): 237–56.
21. *New York Times*, 10 January 1990.
22. *Die Welt* (Bonn), 17 December 1989.
23. See Michael Sobelman, "Poland: The Ecological Tragedy of Upper Silesia," *Environmental Policy Review* 3, no. 2 (July 1989): 28–33.
24. See ibid., 32; and Michael Sobelman, "New Objectives in the Area of Environmental Protection in Poland," *Environmental Policy Review* 3, no. 1 (January 1989): 22–27.

.

I

THE POLITICAL-ECONOMIC SETTING

1

ENVIRONMENTAL POLITICS IN EASTERN EUROPE IN THE 1980s

Barbara Jancar-Webster

Much of Eastern Europe today is fast approaching the so-called ecological barrier, where environmental degradation becomes a roadblock to further economic development. The devastation has reached such proportions that it can no longer be concealed. In those countries with the greatest ecological disruption—the German Democratic Republic (GDR), Poland, and Czechoslovakia—it has become a frequent and urgent topic in the press, but it is now also a common subject in the mass media of all the communist one-party states. In the countries with the worst environmental deterioration, environmental monitoring agencies, expert input, and public participation have gradually become instruments for controlling the approaching disaster.

Before the late 1970s, the environment was primarily the concern of a relatively small number of dedicated natural and social scientists who spoke out at comparatively obscure congresses and meetings and published their findings in the specialized press. By the end of the decade, most of the countries had passed legislation regulating the use of air, water, and soil. By 1980, Poland had enacted more than thirty laws and resolutions.[1] Czechoslovakia had fifty-three laws and regulations,[2] and Yugoslavia more than three hundred.[3] The GDR in 1970, Romania in 1974, and Hungary in 1976 passed comprehensive environmental laws, and in 1977 Bulgaria published revised environmental management guidelines.[4] But these acts, more symbolic than real, were incapable of stopping the advancing pollution.

New forces for environmental protection gathered momentum at the beginning of the 1980s, however, notably in Poland during the Solidarity period. Gorbachev's call for economic and democratic reform and the Chernobyl disaster have accelerated this process. Today in most of the East European countries it is possible to identify an environmental "lobby," organized and led by experts and rooted in growing popular support, with links to economic, party, and government personnel at all levels of the administration. The entrance of an activist public into the environmental arena marks a qualitative change in the environmental politics of Eastern Europe.

THE STRUCTURE OF ENVIRONMENTAL POLITICS: THE ACTORS

Five groups in the East European body politic are engaged in the game of environmental policy: the socially owned economic institutions, the environmental agencies, party and government leadership, experts and expert organizations, and the public (acting individually or collectively in legal or extralegal environmental organizations). These groups are treated in order in the following sections.

The increasing publicity given to environmental issues, and its growing influence on environmental policy, suggests the emergence of the mass media as a sixth environmental actor. However, although there is no denying the expertise of many journalists, official control over what is published prevents the mass media from being considered a truly separate and independent environmental actor in any country.

The Socially Owned Economic Organizations

The economic enterprises enjoy a privileged relationship with the political leadership in all societies, but particularly in the socialist countries. In the socialist system, industry and agriculture fulfill not one but two functions. The first is to produce for profit as in market economies, and the second is to maintain the socialist regime in power. State ownership of the means of production and centralized planning effectively subordinate all institutions to the ruling group within the Communist party. The party cannot, however, simply seek to maximize its power; it must also have economic results.[5] The official rationale for one-party communist rule is that it can provide more equitable, stable, and goal-directed conditions for economic development than those existing in capitalist economies. While concern for economic growth is driving reform in the Soviet Union and in Eastern Europe, concern for security is raising doubts among the more cautious members of the national leaderships that reform will mean their downfall. Because the economy is required to be both a productive organism and the vehicle by which the ruling group assures its dominant position in society, the economic enterprises have the major input in every environmental and economic policy discussion. If there is a question of a trade-off between the economy and the environment, the economy is the beneficiary.

The requirement that the economies of communist one-party states perform these two functions means that the relationship between industry and political power is more interdependent here than in free market societies. To oversimplify, in the free market societies, industry's drive for profit urges industry to political action to pass legislation that will protect and promote its profit-making capabilities. Here, economic interest fuels the impulse to power. In the socialist one-party states, the reverse occurs: the impulse to power fuels the economic machine.

The party-industry alliance is embedded in East European institutional arrangements. The economic organizations are guaranteed access to power on a hierarchical basis depending on their role in promoting and maintaining the interests of the party leadership. The ministries responsible for foreign affairs, national defense, and internal affairs are the most common source of ministerial members for the party politburos of the East European countries. The economic ministries oversee the strategic industrial branches and, less often, mining and agriculture. With few exceptions, the state planning chairman holds the post of deputy prime minister.[6]

In the more centrally managed states—the GDR, Czechoslovakia, Bulgaria, and Romania—the strategic industries have their own separate ministerial organizations, while the consumer industries tend to be grouped under one ministry. The hierarchical order of economic institutions is perhaps most obvious in Czechoslovakia, where the strategic industries (engineering, metallurgy, fuel and power, and electronics) have federal ministerial status, while the lesser ones (forestry, water management, and the consumer industry) are organized at the republican level. In Hungary, heavy industry has retained its centralized direction, while the consumer industries have been largely decentralized and many consumer services privatized. In Yugoslavia, all economic units are self-managing. The economic interests, however, enjoy special representation all the way up the legislative ladder as well as individual self-management associations that, although most active at the republican level, reach up to the federal government.

The economic organizations' position is further privileged by the existence of censorship. Censorship may be supervised by the top political leadership, but it is directed by the ministries and state agencies, which select the information to be released. Although piecemeal information about environmental conditions has become increasingly common in the East European mass media, the global picture is not presented and there is no free public discussion. The editor in chief of *Naší přírodou* (*Our Nature*), the monthly publication of the Czech Society of Defenders of Nature, may assert that he can publish anything he thinks appropriate, but legally and in practice he is bound to observe the instructions given to the Czech Ministry of Health in 1982 standardizing the procedures for authorizing texts for publication at home and abroad. Topics on which information may not be published include pollution of the environment, quantitative data on levels of ionizing radiation, and the incidence of certain defects and diseases, "in particular if this concerns retardation of development of children in some areas with a high level of emissions."[7]

The severity of censorship of environmental information varies from country to country. Before 1980 there was almost no published information on the environment in Poland. The lifting of censorship on environmental data during the Solidarity period immediately unleashed enormous public interest in environmental protection, an interest that has continued to this day. Although martial law ended the Solidarity experiment, Polish scientists still may publish more openly than their Czechoslovak counterparts. But in no communist one-party state are scientists or journalists free to publish anything they want. The silence surrounding data regarding health helps to keep overall public concern low and prevents experts from making public their strongest case for environmental controls—the impact of pollution on health. The general lack of information on radiation levels during the Chernobyl accident in all the East European countries is a case in point.

The Environmental Agencies

The environmental agencies are relative newcomers to the East European political scene, a product of the growing corpus of environmental legislation. In delegating functions related to implementation, these regulations have either relied on existing organizations to manage the environment or created new ones. Where existing ministries and committees (such as agriculture, forestry, and water management) have assumed the management of the environment, the environmental component has been undermined by their traditional orientation toward production. The newly created agencies—the ministries of environmental protection and the committees and councils attached to the national governments—have so far been unable to gain sufficient power to offset the structural advantage of the economic organizations.

To date, only three countries have elevated environmental protection to ministerial status: Poland, the GDR, and, most recently, Hungary. In Poland, environmental protection was the province of the Ministry of Administration and the Local Economy until 1985, when a separate environmental ministry was established. The GDR is the only country where the environmental minister is also a deputy of the Council of Ministers. In Czechoslovakia, Bulgaria, and Romania, committees (councils) on environmental protection are attached to the governments. (In Czechoslovakia these committees are at the republican rather than the federal level and are thus not in a position to influence critical national decisions.) In Yugoslavia an advisory Federal Council on Territorial Management and Environmental Protection has corollary councils at the republican level.

Of the many weaknesses of the environmental agencies, the following three stand out as perhaps the most detrimental to the effective management of environmental protection. First, as in many other countries that have a policy of environmental protection, the bodies of environmental administration tend to have authority for a single area although they themselves are disaggregated and dispersed throughout the government administration. Even in those states that have environmental ministries, the immediate management of pollution control is the responsibility of other institutions. The primary task of the central ministry is to coordinate the work of these institutions to produce a comprehensive program of environmental management. In the remaining socialist states, the environmental committees (councils) attached to the governments are advisory and not administrative. By the early 1980s an environmental department had been created in the planning commissions of all the countries, but in each case its role in the planning process has remained essentially secondary. No country, whatever its record in implementing legislation, has integrated ecological requirements into economic plans. Each economic branch or organization has been required to prepare its own environmental plan, which is transmitted up through the economic hierarchy. Only then are the different subsections of the economic sector's environmental plan (air, water, soil) submitted to the environmental agency charged with coordinating environmental controls in its particular environmental area. The result is to reduce the role of the environmental agencies in both the formulation and implementation of economic plans.

In Czechoslovakia and Yugoslavia the influence of the environmental agencies is further hindered by the peculiar relation of republican and federal institutions. Czechoslovakia has no federal environmental protection agency. The functions of environmental management are assigned to the ministries of health, construction, and forest and water conservation at the republican level. Nature conservation is the province of the two republican ministries of culture, which manage the national parks and nature reservations. The parceling of Czechoslovakia's environmental protection has spurred experts to demand a comprehensive environmental protection law at the federal level and the establishment of a federal environmental protection agency. The Institute of State and Law of the Czech Academy of Sciences is currently drafting such a law.

In Yugoslavia virtually all power resides at the republican level, and any federal compacts that come into being are made through agreement among republican institutions. The economically more advanced and wealthier republics, notably Slovenia, have stricter environmental regulations than the less advanced republics and have created the institutional machinery to enforce them more efficiently. The Slovenes argue that the standardization of environmental norms would either financially penalize the less developed republics by their severity or be too lax to satisfy the more environmentally demanding standards of the more developed republics. Although this argument has some merit, the lack of standardization of norms and laws across the republics and the persistence of interrepublican rivalry have seriously impeded the realization of a comprehensive environmental program. This can be seen, for instance, in the long-standing problem of cleaning up the Sava River.[8]

Second, the system of fining violators of environmental laws undermines the power of environmental agencies. Although state environmental inspectorates may cite an enterprise for violation or demand that a factory be closed, the disposition of the case rests with local government. Local authorities may be reluctant to enforce the law because of local concern for production, jobs, and community well-being, or because of pressure from the central party and government administration; ministries, for their part, are increasingly reported in the East European press to deliberately ignore local environmental complaints, leaving the enterprises under their control free to continue to pollute. In other cases, the enterprise may prefer to pay the fine (and fines can be quite high), because fines are programmed into enterprises' operating expenses and do not affect profits or other indicators of success. The retroactive adoption of pollution controls, on the other hand, characteristically involves great expense. In the early 1980s, most of the countries of the bloc promulgated regulations ending the practice of calculating pollution fines as operating expenses, but experts doubt whether the new regulations will have much effect. There are too many ways in which fines can be hidden in an enterprise's budget.

Finally, financial and technological problems complicate the work of the environmental agencies. All the East European environmental programs are chronically underfunded. Equally important, all the East European countries are dependent on pollution control technology imported from the West. Importation means spending scarce hard currency, which the governments are reluctant to do for environmental protection when it is needed to pay for energy, spare parts, and other technological imports. Moreover, because it tends to be newer, environmentally safe technology costs more than older equipment. In Yugoslavia, where firms have more independence in ordering abroad, the federal government has suspended the import tax for pollution control equipment. Many enterprises have been publicly criticized, however, for choosing older, environmentally destructive equipment in place of the more expensive, newer technology.[9]

The environmental agencies' weak structural relationship to the political leadership augurs poorly for their assuming a major role in environmental policy-making. There is little they can do without outside support. Among the most vigorous advocates of agency efficiency have been experts working in the academies of sciences and other research institutes not formally associated with the environmental administrative bodies. Indeed, in all the East European countries, the experts' support has been one of the key factors in the adopting of regulations and legislation to strengthen these bodies' powers to monitor and enforce.

Party and Government Leadership

By its power to establish policy priorities, the central party and government leadership is the decisive actor in environmental matters, as in all other areas of policy. Until recently, environmental protection was a virtual nonissue in all East European countries. The party leaderships' reluctance to abandon Stalinist production-oriented policies is rooted in their dependence on the production and control functions of the economy to remain in power. This reluctance is reinforced by the advanced age of the leaderships coupled with the relative economic backwardness of many of the countries. With the exception of

Yugoslavia, the current East European leaders are first-generation postwar Communists. Immediately upon assuming power, the new leaders were determined to secure their positions and direct the recovery of their economies from the destruction of World War II while simultaneously adjusting to the harsh exigencies of the cold war. The more economically backward Balkan countries faced the additional problems of moving out from the shadow of several centuries of Turkish rule and embarking on rapid industrialization. Little thought was given to the environment at that time, and the focus has remained on economic development. One of the tasks of the environmental experts in all the countries has been to convince this aging leadership of the urgency and relevance of environmental problems.[10]

The situation is somewhat different at the local level, where the burden of enforcement has fallen on the local territorial units. Because the local communities suffer directly from the effects of pollution, either in increased costs of health care or reduced economic potential, the local East European party and government officials are probably more sensitive to the need for environmental control than are the national leaders. In her ongoing study of public attitudes in local communities in Poland, Siemienska found that between 1978 and 1984, awareness of the environment had increased among local government officials. In 1978, 96 percent of a sample population of local officials identified protection of the environment as a "major" issue, placing the environment third after economic and human rights.[11] The constructive roles played by the mayor of Cracow, Józef Gajewicz, and the environmental director of the Cracow voivodship (province), Wojciech Szczepanski, in the closing of the Skawina aluminum plant in December 1980 lend credence to Siemienska's findings.[12]

Although a concerned local political administration can certainly help to improve local environmental conditions, its options are highly circumscribed. First, it can do little against the power of a central ministry. Second, although local government can and does support air, water, and soil conservation programs, major improvements are often technically beyond its powers. Third, most environmental measures also cost a great deal of money, which in centrally managed economies must come from the central government. But the money allotted for environmental protection is insufficient and frequently is not spent at all or is deflected to other purposes.[13]

Local officials are under particularly difficult restraints in Yugoslavia. In the self-management system, the community is required to fund all its own projects. If it is to undertake large-scale environmental programs, it must appeal to local industry and local citizens. If the citizens' money is used, there must be a referendum. To their credit, Yugoslavs have frequently voted for increased taxes to protect the environment, notably in cleaning Lake Palić in Vojvodina, developing a water purification system for Lake Ohrid in Macedonia, and building a clean water system in Sarajevo. Industry has also donated funds, as in the case of the Zastava automobile plant in Kragujevac and the fruit-processing industries in Valjevo. But not every community has an environmentally concerned citizenry or a major industry to fund its needs. In that case outside money must be sought, either at the republican and federal levels or from international sources.[14]

In all the East European countries the local environmentally conscious officials are limited by the power of the major economic enterprises in their areas. The commanding influence of the economic enterprises at the local level is in turn strengthened by the need of the central party and government functionaries to ensure economic growth in such a way as to keep themselves in power. Until this circular relationship is broken, environmental protection will not receive the attention the deteriorating situation merits.

Experts and Expert Organizations

The experts have been able to create a more independent position for themselves in environmental politics than any of the other environmental actors. It is generally recognized that experts play a role in policy-making in direct proportion to a society's level of modernization.[15] Wolchik's research substantiated the ever-increasing importance Soviet and East European scholars attribute to the harnessing of "the scientific-technological revolution" to economic development.[16] In this respect environmental experts may enjoy a greater confidence among decision makers than other scientists. Policymakers appear to accept more readily the intervention of specialists in solving environmental problems than in other areas.[17]

The need of their participation has put the experts in an unparalleled position to influence policy-making. Pioneering studies on the nature of this influence have been done by Skilling, Kelley, Solomon, and Lowenhardt.[18] In his study of changes in Soviet policies on land and water management, Gustafson presented considerable evidence for the major role specialists played in influencing a shift in Brezhnev's thinking about the relationship of the environment (air and water) to economics.[19] Research has confirmed the key role played by experts in modifying leaders' understanding of the environmental issue.

Experts are required at every stage of environmental management.

The natural scientists put environmental protection on the communist legislative and regulatory agenda. Economists, jurists, social scientists, writers, and journalists have all done their part in pushing back the curtain of censorship, publicizing the extent of damage, and elaborating the legal and technical parameters for solutions. Since the early sixties at least, the leadership has had access to increasingly sophisticated quantities of confidential statistical data. Though it might well be argued that the ruling group has not reacted as forcefully as the situation demanded, the fact remains that there now exists a network of environmental legislation in all the East European countries, drafted and enacted at the instigation of the party leadership, in partial fulfillment of the experts' environmental agenda.

Experts on legislative committees and in expert institutions draft the environmental legislation. Experts carry the discussion of the draft law through the various stages of the legislative process. Experts hold seats on the legislative committees empowered to make the necessary recommendations. Experts in the environmental agencies implement the law after it is passed. Implementation implies a gamut of activities, including the development and application of technology and monitoring systems to control pollution, the education of elites and the general population as to the importance of environmental protection, control and inspection, generating feedback on the progress of environmental measures, and making suggestions for their improvement.

Experts play the principal role in the public dissemination of information about the environment through the mass media. Although this role, together with the importance of specialists in the environmental and economic policy process, gives them an opportunity unique in the socialist system to argue for policy options not previously endorsed by the leadership, they are constrained by their employee-employer relationship to the central power structures of the party and economic ministries. The proliferation of environmental regulations has increasingly institutionalized this relationship. The authority of environmental experts depends on their position in a particular bureaucratic structure. Research is compartmentalized within a multiplicity of institutes administered by different bureaucracies. Expertise is thus subject to fragmentation, duplication of effort, and bureaucratic jealousies where interdisciplinary cooperation is required. Moreover, the bureaucratic control of information focuses specialists on narrow research projects, severely inhibiting the comprehensive study of environmental issues. In Yugoslavia the influence of specialists is further weakened by the need for each research institute to be self-managing, the absence of a solid domestic environmental research program, and the consequent need to seek scientific and technological answers abroad.

The Public

The fifth environmental actor is the public at large. In no Communist country is the public allowed to organize and express itself on its own initiative. In Yugoslavia, although individuals and groups have considerable freedom to express their views at the local level, there exists no environmental organization at the republican level. The chief mobilizer of public environmental interest is the Federal Council for the Protection and Improvement of the Environment organized under the aegis of the Socialist Alliance, which unites all professional and mass organizations in the country. The role of the council in the public domain is to propose environmental actions which, after formal adoption, are implemented by local government through the mass participation of interested citizens.

Mass environmental organizations exist in the other East European countries as well, and they enjoy varying degrees of public acceptance. Poland's League for the Preservation of Nature has a one-hundred-year-old tradition and over a million members, but it is not viewed as very effective either at home or abroad.[20] The Society for Nature and the Environment, established formally in the GDR in 1980, has been unable either to harness citizen concern or to counter the growth of independent and church-associated environmental groups. In Czechoslovakia, where each republic has its own organization, the Slovak society was organized in 1973 and the Czech union in 1979. In Hungary, where the Association for the Protection of Birds has been in existence since the end of the last century, and the more recent Association of Friends of Nature claims a hundred thousand members actively engaged in nature protection and conservation, perhaps the broadest public is reached through the environmental committees attached to local units of the Patriotic People's Front. Bearing some similarity to the Yugoslav councils for the protection and improvement of the environment, these committees mobilize citizens at the local level to conserve and study nature.[21]

The role of these officially sanctioned mass organizations is to "cooperate" with the authorities within the guidelines of the national economic and social plans, not to develop grass-roots views or initiatives on environmental issues. Each organization is established through enabling legislation that assigns it important educational, propaganda, and civic actions.[22] Each is allowed its own journal as well as access to the local press, and each holds seminars, "environmental universities," and special education sessions for party and government elites and the public. These organizations recruit volunteers for reforestation and beautification projects, trail marking, wildlife preservation, and scientific data gathering. One of their principal functions is to make young people aware of the need for environmental management. Among the most publicized projects of the Hungarian Patriotic People's Front, for example, is the annual environmental protection poster contest for schoolchildren.

Educational work among young people is also done by the Communist youth organizations or specialized environmental youth groups working with the youth organizations. In Czechoslovakia, young people up to the age of fifteen may enroll in Brontosaurus, an organization attached to the Czechoslovak League of Youth. In Yugoslavia, the Gorani Movement is known throughout the country for its recruitment of young people for the much-needed reforestation program. The Association of Young Researchers brings together university students and has a strong environmental section. In cooperation with the Yugoslav League of Communist Youth, it organizes research programs for young environmentalists during the summer months. All over Eastern Europe, students assist researchers and academics in collecting data on nature preserves, the migratory habits of wild animals, the pollution of rivers and lakes, and the prevalence of endangered species. Young people in the Czech Society of Defenders of Nature take care of a herd of about sixty Hucul horses, an endangered breed native to the Carpathians.[23] In Czechoslovakia and Yugoslavia, as in Poland, Hungary, and the GDR, the limits to independent initiative may discourage participation in such organizations, but they present opportunities for political involvement otherwise hard to obtain.

The five actors discussed in the preceding sections interact in dynamic and fluid ways. As the experts and environmental agencies strive to expand their regulatory and consultative power, they seek support from proenvironmental party and government leaders at both the regional and national levels. At the same time, the experts and environmental officials need a popular constituency to legitimate their activities. The specialists lead and advise the environmental organizations, and the knowledge they impart to the public determines in large measure the scope of popular activity. The experts' success in translating their sense of urgency into effective policy depends both on the relevance of a particular environmental issue to their political or economic employers and on their ability to direct public interest toward areas that are acceptable to party and government authorities. Although the environmental agencies need the public to report and protest violations of the law in order to maximize their regulatory powers, they have an interest in restricting popular concern to local issues lest it get out of hand and mount a serious challenge to their authority on a national scale. Although both the environmental agencies and the experts must be careful, they are obliged to court the public in their efforts to offset the permanent weight of vested state economic interests.

EAST EUROPEAN ENVIRONMENTAL POLITICS IN THE 1980s

Although vested state economic interests appear to be the inevitable winners in the game of environmental policy in Eastern Europe, the regimes' recognition of the need to address the deterioration of water, air, and soil quality has given the other environmental actors new opportunities to influence environmental policy. Five recent developments have modified the positions of the five actors: the official adoption in all countries of a variety of new measures to combat pollution, the emergence of a legitimate expert environmental lobby to counter official environmental pronouncements, the persistence of the dissident movement, the growth of independent popular environmental groups, and the challenge posed by the ascendancy of Gorbachev in the Soviet Union.

Innovation in Official Policy

Innovation in official policy may be seen in two areas: the tightening of domestic rules and regulations controlling pollution, and bilateral and multilateral cooperation with neighboring countries in both the Eastern and Western blocs. On the domestic level, the 1980s saw a flurry of legislative activity in several of the East European countries. Those most affected by pollution have made progress in integrating environmental policy with economic policy. Some developments have already been mentioned, but others should be noted here.

Most East European countries doubled state investment in the environment in their 1986–90 plans. Hungary published a document in 1980 entitled *Conception and Requirement System for Environmental Protection in Hungary*, making environmental protection an integral part of national economic plans.[24] It has given rise over the past two years to a series of new regulations putting more teeth into the Hungarian environmental program. Czechoslovakia adopted an ecological rehabilitation program in 1985, which for the first time addressed environmental protection comprehensively at the central government level and called for doubling the investment in pollution controls for the 1986–90 plan. A new law passed in 1986 required that any institution planning new development send its plans together with something akin to an environmental impact statement to the responsible environmental organs for review.[25] In Poland, where more than a third of the population lives in regions threatened with ecological disaster (amounting to 11.3 percent of Polish territory), the program adopted by the Tenth Party Congress in fall 1986 recognized that environmental devastation had become a barrier to further socioeconomic development and gave environmental protection the same priority as housing, food, and energy. The current five-year plan assigns 3.3 percent of investment funds to environmental protection.[26] The environment also figured prominently on the East German agenda at the Eleventh Socialist Unity Party (SED) Congress in March 1987. The Bulgarians are currently engaged in drafting comprehensive environmental legislation to replace the outdated laws of the 1960s.[27] Yugoslavia incorporated environmental protection into its 1981–85 plan, requiring that all new development be supported by bank loans as necessary to comply with mandated pollution controls.

Further evidence of governmental innovation may be found in the recent movement toward cooperation between neighboring countries as transboundary pollution has become an important source of international friction. The most controversial issues involve transboundary air pollution (e.g., the impact of acid rain on the border areas of Poland, the GDR, and Czechoslovakia; and of chlorine pollution from Romania across the border into Bulgaria) and the pollution of national waterways and rivers crossing international frontiers (notably the Oder, the Tisza, and the Danube region where Bulgaria has proposed a Balkan ecological pact and an intergovernmental convention on protecting the environment).[28] Within the Council for Mutual Economic Assistance (CMEA), the GDR, Poland, and Czechoslovakia have set up working agreements on the control of transboundary air and water pollution in the border areas. The Soviet Union, Hungary, Yugoslavia, Czechoslovakia, and Romania signed an agreement in 1986 on maintaining the high quality of water in the Tisza River.[29] In addition, an increasing number of environmental conferences organized by CMEA have been devoted to topics of mutual concern.

All the countries except Romania have actively participated in these meetings, and recent signs indicate that Romania may be on the verge of joining.

An even more important indication of the seriousness with which the East European leaderships are beginning to take environmental pollution is the new cooperation with countries in Western Europe. Hungary was the first to conclude a bilateral working agreement with a Western country—Austria,[30] which ironically now provides credit for the Hungarian share in the construction of the controversial Gabčikovo-Nagymáros Dam, a hydroelectric power and navigation system on the Bratislava-Budapest section of the Danube. Czechoslovak relations with Austria are less cordial, partly because of an earlier dispute over planned joint construction of another dam on the Danube. The project was delayed in the face of official Czechoslovak objections to the location of the dam and strong Austrian popular protest about its potential ecological impact.[31] The GDR is currently holding discussions with the Federal Republic of Germany (FRG) with a view to cooperative control in transboundary air and water pollution along the Elbe. The GDR has made fresh approaches since Chernobyl to the FRG for assistance in assuring the safety of nuclear reactors.[32] And in 1988 Czechoslovakia made a major commitment to control pollution by signing an agreement with the FRG for the purchase of scrubbers for its coal-fired electric plants.

In the area of East-West regional cooperation, Czechoslovakia, Hungary, Bulgaria, and Yugoslavia have signed the United Nations Economic Commission for Europe Convention on the Long-Term Pollution of the Atmosphere. The riparian nations are party to the interstate agreement on pollution control of the Danube, and Yugoslavia actively participates in conferences and studies held in connection with the protection of the Adriatic and Mediterranean seas. All the East European states (including Romania at the last minute) sent representatives to the first East-West meeting, the Global Possible Conference on the Environment, which took place in Washington, D.C., in June 1984. Romania was not present, however, at the International Conference on Environmental Management in Versailles in 1985. On a global level, all East European countries participate in the United Nations environmental programs, although some are more active than others; Hungary, in particular, values international environmental cooperation as a source of scientific information and technological assistance.

Although East European officials tend to downplay the event, there is no doubt that Chernobyl played a large role in putting further pressure on their regimes to take stock of the environment. No press was more open in its official reporting of the disaster than the Polish press, although even in Poland there were large gaps and silences subsequently challenged by underground publications.[33] Reporting in Hungary prudently followed the Soviet lead, even though Hungary was to be the only East European country to pay partial compensation for Chernobyl-related losses.[34] Although Polish officials stated publicly that "the accident would have no effect on the realization of Poland's nuclear power program,"[35] Chernobyl pushed the GDR toward closer cooperation with West Germany in nuclear safety and forced the Yugoslav federal government to abandon its plans for building a second nuclear power plant. The particularly strong negative impact from Chernobyl that Romania felt in its agricultural exports may help to explain its subsequent willingness to engage in international environmental cooperation.[36]

The Expert Lobby

The experts have been the chief beneficiaries of renewed official activism in the environmental area. In the 1980s they emerged as the most visible and authoritative spokespersons for the environment. Two developments merit mention.

First, experts have secured institutional bases around which has formed an unofficial but legitimate environmental scientific lobby. It is held together by the informal ties of an "invisible college" of experts who, partly because of the relative novelty of the environment as a field of study, tend to know one another personally. Acquaintance typically holds over from university days and extends beyond the domain of natural sciences into the social sciences. Official opportunities for interdisciplinary exchange are provided through environmental conferences and contact in the professional and mass organizations of the National Front or corresponding institutions. In Czechoslovakia, the Presidium of the Czech Society of Defenders of Nature is composed of ranking experts from the economic, legal, and natural sciences. The composition of the presidium is not an accident but the result of long years of contacts and collaboration among its members. In Yugoslavia, contacts are made through the Socialist Alliance. Although republican autonomy and language barriers tend to confine specialists within their own republics, opposition to the Tara River Dam in 1985 indicated that it was possible, through the Socialist Alliance, to generate a nationwide expert response on short notice to a perceived environmental threat.[37]

Second, in all the countries, the experts' exasperation with the pace and efficacy of pollution control has increasingly distanced them from official policies. In Czechoslovakia, most of the expert lobby has remained low-key and submissive to government dictates, but there are signs that middle- and low-ranking experts are taking a more independent stance than their superiors. A recent publication on the environment, *Životní prostředí: věc veřejná i soukromá* (*Environmental Protection: A Public and Private Matter*), is the product of collaboration by a group of younger experts from most of the institutes engaged in environmental research. Although it is not openly critical of the party-industry alliance, it urges that current anthropocentric policies be abandoned in favor of a more holistic approach to man's relation to nature, using Western examples to illustrate the effectiveness of this approach.[38] (In Hungary, to relate a personal anecdote, a young scientist categorically informed me that because traditional Marxism-Leninism posits man's triumph over nature, to be proenvironmental is to be fundamentally antiregime.)

Experts have increasingly made their position public. In Yugoslavia, the coordinated action of environmental experts to stave off construction of the Tara River Dam was publicized by reports, statements, and letters in the press. In Yugoslavia, a Belgrade radio program that not only provides information on pollution but actually exposes polluters has reportedly become one of the most popular in the city. The growth of expert opposition to the regime's environmental management in Poland may be measured by the Sejm's rejection in January 1987 of the government's report on environmental pollution. The Sejm criticized the report as "unreliable, contradictory, too general, and overoptimistic."[39]

The experts' tenacity in holding their ground in public against the party-industry alliance has been reinforced by the emergence of an extralegal lobby operating outside authorized channels but with links to officially recognized institutions. Because of the official endorsement of environmental protection, this lobby cannot really be considered dissident, although it is supported by dissidents. Using both official and underground media, the experts are able to mobilize domestic and international opinion to modify the regime's policy. Two recent examples illustrate this point.

By 1987, chlorine pollution from a Romanian chemical plant across the Bulgarian border had assumed levels dangerous to human health in the Bulgarian town of Ruse. Talks between Romanian and Bulgarian authorities did not end the problem. In December of that year, to induce the national authorities to put pressure on the Romanians to close down the plant, the Union of Bulgarian Artists organized an art exhibit in Ruse featuring macabre grey and yellow paintings of dying birds, rotting fish, dead trees, and children wearing gas masks. The paintings were accompanied by grim statistics provided by the medical profession on the extent and effects of chlorine pollution in the area. Taking advantage of the nationwide publicity given to the exhibit, the union urged all intellectuals to join in the fight against pollution.[40]

At the end of 1986, the Křivoklat Nature Reserve authorities in Czechoslovakia discovered that plans were being made to build a dam over the Berounka River near Prague, in the heart of an area under the protection of the Man and the Biosphere project of the United Nations Educational, Scientific, and Cultural Organization (UNESCO). The park authorities publicly protested the proposal and asked the Czech Society of Defenders of Nature to publish an article on the subject, which subsequently appeared in the union's monthly, *Naší přírodou.*[41] Through international environmental sources it was discovered that construction of a dam on the Tara River in Yugoslavia, another UNESCO biosphere, had been stopped. Less than a year later, following the completion of an environmental impact study in fall 1987, the Czech authorities halted the Křivoklat project.

An entirely different reception was given by the Slovak authorities to the independent report on pollution in the Slovak capital of Bratislava released in June 1987 by a group of scientific and medical experts associated with the official Slovak Society of Defenders of Nature and the Countryside. Entitled *Bratislava nahlas* (*Bratislava Aloud*), the report described the ecological degradation of the city and criticized the urban planning that had drastically brought down the city's quality of life and given Bratislava the highest suicide rate in Czechoslovakia. The authorities responded to the report's publication in the underground press by accusing its authors of exploiting environmental issues to promote their own antistate agenda.[42]

The Institutionalization of Dissidence

The third development of the 1980s with important implications for environmental policy in Eastern Europe has been the institutionalization of dissidence. Although Solidarity and Charter 77 are widely different in composition and aims, each in its own way has become what might be called an extralegal loyal opposition. Neither overtly seeks the overthrow of the existing system, but each demands reforms so far unacceptable to the ruling group. Despite the imposition of martial law in Poland and the ongoing persecution of Charter 77 signatories in Czechoslovakia, neither group has disappeared. Solidarity took up the cause of the environment in Poland, while Charter 77 has played the role of the environmental conscience of Czechoslovakia for more than nine years.[43]

In both cases an underground press has provided access not just to a national audience but to an international audience as well. For the past decade, Charter 77 has published a sixty-page, typewritten, monthly bulletin, *Informace o Chartě*. It has also published more than three hundred statements on human rights particularly relevant to Czechoslovak life and society and several major statements on the environment. Charter 77 responded to the authorities' handling of information about the Chernobyl accident with an open letter to the Czechoslovak government and National Assembly, demanding full publication of the facts.[44] Its tenth anniversary statement, "A Word to Fellow Citizens," signed by thirteen hundred persons, unambiguously called on Czechs and Slovaks to acknowledge their responsibility to future generations by engaging in public discussions about the environment.[45] And in April 1987, the Chartists published their second full-length document on the environment urging a restructuring of the economy and the elevation of environmental protection to the status of government's "first priority."[46]

The liberalization of the official Polish press in 1980–81 played a major role in the closing of the Skawina aluminum mill in Cracow. Since then, censorship has again been more strictly enforced, but an active and prolific underground press has kept the Polish population perhaps better informed of the true state of internal environmental conditions than any other East European public. Solidarity's success in mobilizing the support of scientists, which has enabled the underground press to counter official reports with more accurate information about the effects of Chernobyl, is indicative of the degree of the movement's expertise and organization.[47] The institutionalization of a "loyal opposition"[48] has thus created a second permanent source of information about the environment and, still more important, makes available to the public alternative environmental options that are more rational than those offered by the authorities.

The Rise of Independent Environmental Groups

The growth of the independent grass-roots environmental groups in Eastern Europe is one of the most fascinating developments of the 1980s. Experts have assumed the initiative in forming and leading these groups, which are mushrooming in every country in Eastern Europe with the possible exceptions of Romania and Albania. In East Germany, an independent environmental movement publishing a *samizdat* (clandestine) environmental journal, *Umweltblätter* (*Environmental Pages*), has taken off under the aegis of the Lutheran church. Radio Free Europe estimates that in Poland there are more than two thousand organizations active in environmental protection.[49] Their precursor is the Polish Ecological Club, formally organized in September 1980 following the Gdańsk agreement.[50] It began publishing a popular monthly and soon claimed a membership of about a thousand. Under its sponsorship the Polish people learned for the first time of the seriousness of pollution in their country. The club was the moving force behind the drive to close the polluting Skawina aluminum plant, and its experts prepared the memorandum demanding an end to all processes emitting fluorine and hydrogen fluoride.[51] With the imposition of martial law in December 1981, the club was proscribed and went underground, resurfacing after Chernobyl.

The Chernobyl accident was a special stimulus to the expansion of the independent groups. In Białystok, Wrocław, and Cracow, thousands of workers signed protests against atomic energy. In May 1986, the Independent Ecological Committee was founded in Lublin with the aim of regularly informing the public of environmental threats in the Lublin area. Also in May 1986, two health and science committees linked to Solidarity formed a working group for the protection of the environment. Its aim was also to inform the public through a publication entitled *Zagrożenie* (*The Threat*). In an open letter, the Polish Ecological Club appealed to the Sejm to place atomic energy under public control, asking that the construction of all new plants be subject to national referendum. As in the GDR, the Church (this time the Catholic church) has also decided to lend its considerable prestige and power to the environmental cause by sponsoring a new organization, appropriately named the Saint Francis of Assisi Ecological Movement.[52] Significantly, many of the new groups have attributed the Chernobyl accident directly to the Soviet political and economic system, claiming that the current state of the environment in Poland is the result of the regime's refusal to listen to public opinion.

Nuclear power is not the only environmental issue of concern to the Polish public. In January 1987, the Polish nightly television news carried a report about ecological groups and students demonstrating in Wroclaw against the accidental discharge of heating oil into a tributary of the Oder by a Czechoslovak lime and cement factory in Ostrava-Kunčice in northern Moravia. The Czechoslovak government's delay in reporting the November 1986 accident by three days led the Polish government to protest what it characterized as inadequate and inaccurate information and to demand compensation. The Wroclaw demonstrators claimed that the cleanup had not been properly carried out and that spill-related pollution of the Oder was continuing.

The incident serves to illustrate the potential force of the extralegal environmental movement. Although none of the Polish groups is currently strong enough by itself to mount a serious challenge to official ecological policy, the Sejm's rejection the same month of the regime's proposed environmental program implies widespread support for their protests and suggests their latent power. The regime is conscious and fearful of this power. Wishing to neutralize the movement, it has tried without success to merge the independent groups with the official Nature Defense League or to unite them into a new mass organization under the Patriotic Movement for National Rebirth. But the future integration of the Polish environmental movement appears to lie in another direction. In May 1987, ecological groups participated in the first conference organized unofficially in Poland, on "International Peace and the Helsinki Agreements." Organized by Freedom and Peace, a group of young peace activists, the meeting brought together young people from both Eastern and Western Europe to discuss the Helsinki agreements, peace, human rights, nonviolence, and post-Chernobyl environmental issues.[53] Hungary's unofficial groups have emerged largely as a result of the controversy over the joint construction with Slovakia of the Gabčikovo-Nagymáros Dam. The project was conceived in the 1950s, but current operations date from a 1977 interstate agreement setting forth the obligations of each country and pushing back the completion date from the mid-eighties to 1993.

Almost from the moment the interstate agreement was signed, the proposed project was attacked in the Hungarian mass media. A group calling itself the Association for the Protection of the Danube Region requested permission to organize but was unable to obtain the requisite official permit. In 1983 an outspoken critic of the proposed dam, the biologist János Vargha, formed Hungary's first independent ecological group, the Danube Circle. A second group calling itself the Blues emerged in 1984. The Blues drew international attention in 1985, when they sent a letter of protest to the Hungarian National Assembly and the Council of Ministers, which was signed by fifty leading cultural and scientific figures. It was later circulated among the general public and signed by ten thousand people (only six thousand signatures accompanied the letter, because the rest had been seized by the police).[54] In fall 1985, the Blues issued a new appeal to the public to protest the construction of the dam after the government had made public its belated study of the project's ecological impact.[55] In 1986, the Danube Circle was awarded the "alternative Nobel Prize" awarded by the London-based Right Livelihood Award Foundation. Thoroughly embarrassed, the Hungarian government permitted only Vargha to accept the prize. The Danube Circle insisted that it was not seeking a confrontation and invited the Hungarian Academy of Sciences to supervise the use of the prize money; the officials rejected the offer. By this time sufficient public awareness of problems related to the dam had developed, for some twenty-five hundred signatures were secured for a petition demanding a referendum on it—at that time an extraordinary demand for a Communist country. This request was also rejected by the government.[56]

Perhaps the most significant accomplishment of the Hungarian environmental movement has been in forging international contacts. In December 1984, the Danube Circle appealed directly to the Czechoslovak government to abandon construction of the dam. In 1985, the West German parliamentary Green group led an unofficial delegation to Hungary to protest the dam. During 1986 an extraordinary degree of cooperation developed between the Hungarian groups and the Austrian Greens. A joint meeting of the leaders of the Blues and the Greens resulted in the "Danube Walk" in Budapest on 8 February 1986, which was broken up by the police. When in May 1986 the Austrian and Hungarian governments proceeded to sign an agreement providing Austrian credits for the construction of the dam, the two groups apparently joined forces to try to prevent the agreement from being ratified. In an unheard-of action for any independent group in Eastern Europe, thirty Hungarian intellectuals put a notice in the Austrian daily *Die Presse* formally petitioning the Austrian parliament to review the terms of the agreement.[57] In 1987 the Danube Circle established links with Charter 77, which had also condemned the Gabčikovo-Nagymáros Dam. In March 1988, scientists from the Czechoslovak Academy of Sciences added their voices to the opposition to the dam.[58]

As opposition to the dam grew more vocal and widespread, other local groups emerged protesting Hungary's agreement to dispose of toxic wastes from Austria. In communities where the waste was to be dumped, grass-roots organizations succeeded in forcing local authorities to refuse to accept it. By the end of 1985, the government had to contend with a plethora of independent groups, each with its own programs and tactics. As in East Germany and Poland, the Hungarian environmental movement publishes a *samizdat* journal devoted exclusively to environmental issues, *Vizjel* (*Watermark*).

Environmental activism in Czechoslovakia has been slow in forming. Tis, an unofficial Czech environmental organization launched in the early seventies, was unceremoniously disbanded in 1978 before the establishment of the Czech Society of Defenders of Nature. The absence of more recent independent activity may be attributable in part to the regime's rigid control of the population, and also to the activist image of the approved environmental organizations.[59] The Slovak Society of Defenders of Nature (counterpart of the Czech society) has an extensive program of nature conservation and has taken the lead in publicizing Slovak ecological problems (e.g., the pollution of the Hron River in Central Slovakia). Significantly, the experts who drafted *Bratislava Aloud* relied on the Slovak environmental organization to collect the pertinent data.

In an effort to extend its activities, the Czech Society of Defenders of Nature is seeking to lessen the constraints under which it must wage the environmental battle. At the present time, both the Czech Society and the Slovak Society conduct their activities under the supervision of the republican ministries of culture. Officials of the Czech Society would like to take it out of ministerial control and make it a member of the National Front, an umbrella group of professional organizations.[60] Not only would membership in the National Front give the Czech Society equal status with the other mass organizations, it would legitimate the union's engagement in politics. Within the National Front, the Czech Society could propose or support pro-environmental candidates for elections and play a greater role in the elaboration of the National Front's environmental program—activities currently denied it by the Ministry of Culture. František Hron, president of the Czech Society of Defenders of Nature, is optimistic that membership in the front will come sooner or later. The Czech Society is also attempting to broaden its international contacts. In February 1988, Hron met in Prague with representatives of the Adirondack Mountain Club of upstate New York, the first face-to-face encounter between a U.S. environmental group and an official organization in Czechoslovakia.[61]

New developments suggest that public concern over deteriorating environmental conditions in Czechoslovakia is breaking out of its official constraints. As elsewhere in the bloc, Chernobyl focused public attention on the dangers of nuclear power. Austrian students protested in Prague and Bratislava against Czechoslovakia's construction of nuclear power plants along Austria's border. Young Czechs staging their own smaller demonstration against nuclear power were all immediately arrested.[62] Of special significance has been the grassroots protest in northern Bohemia, the first of its kind in Czechoslovakia. Ecological devastation in the region has forced the authorities to take the protest seriously. In January and February 1987, three hundred people from the Chomutov district (one of ten districts in the northern Bohemian region) signed a letter that was first sent to the chairman of the district government and then to the Czechoslovak prime minister and the presidium of the Czechoslovak Communist party. It is not clear whether the letter complained only about the inadequacy of the warning system when air pollution reached dangerous levels or about the quality of air in the district generally. The matter was taken up at the regional party conference in October 1987. The party regional secretary admitted that the petition's criticisms were justified and called attention to the threat to public health posed by pollution. He also cited the public's lack of faith that the party would do anything to remedy the matter and predicted a breakdown in Czechoslovakia's international credibility if nothing were done to reduce sulfur emissions.[63]

Until recently, environmental activism in Bulgaria and Romania appeared to be confined to experts and intellectuals. The dangerously high levels of chlorine pollution in the Bulgarian town of Ruse noted earlier in this chapter have transformed Bulgarian public opinion. A series of spontaneous popular demonstrations in Ruse in fall 1987 forced the Bulgarian authorities to lift censorship on the issue and to enter into discussion with Romanian officials. Continued demonstrations encouraged the Union of Bulgarian Artists to mount the exhibit that has been mentioned and led the Union of Bulgarian Writers to publish a strongly worded condemnation of both the Bulgarian and the Romanian responses to the situation.[64]

In Yugoslavia, environmental issues have remained largely local. Because the system of enterprise self-management requires that all parties to a project must agree to it before it can be implemented, environmentalists have been given a limited voice in decisions regarding national parks. There have been isolated cases of public protests over the location of polluting industries, most notably the controversy over the building of a nuclear power plant on Vir Island in the Adriatic. The public was also successful in holding up the construction of an offshore plant backed by French and American money near the city of Ulcinj on the Adriatic. On the federal level, the decision not to build a hydroelectric station on the Tara River was most welcome, because it would have flooded one of the most beautiful wild river gorges in Europe. There has been less movement in cleaning up the pollution that threatens the Adriatic and all of the country's major waterways. The Tara itself risks being severely polluted by discharges from a lead smelter above the famous canyon. Interrepublican controversy over pollution of the Sava River on the border between the republics of Croatia and Serbia continues unabated.[65]

The news of Chernobyl broke at a most opportune moment in Yugoslavia, when popular attention was focused on a nationwide debate over construction of four new nuclear power plants. During the April 1986 election campaign for the federal parliament, members of the Communist Youth League in Serbia drew up a petition calling for an end to nuclear energy in the country. Although the young people collected some 120,000 signatures, they were denied access to the media. Their initiative therefore had little effect, and the new government approved the purchase of four nuclear reactors. On 10 May 1986, hundreds of young people under the sponsorship of the Communist Youth League of Slovenia demonstrated in Ljubljana against the risks of polluting the earth and constructing nuclear power plants in Yugoslavia. Serious reservations about the future of nuclear energy were expressed in the nuclear community.[66] In Belgrade, intellectuals formed the Committee for the Protection of Man and the Environment to address nuclear and other environmental issues.[67] Popular anxiety over nuclear power ran so high in Slovenia that a public conference entitled "Ecology, Energy, and Consumption" was organized in 1987 under the sponsorship of the Socialist Alliance. The conference's demand for a moratorium on nuclear power until the year 2000 was approved by the Slovenian legislature. Developments in the nuclear area are exemplary of the interpenetration that has developed between grassroots environmental initiatives and official organizations.

The formation of independent environmental groups and the increased activity of officially sponsored environmental organizations are not insignificant developments. Their presence in East European environmental politics has yet to put the party-industry alliance severely to the test, but the spur to environmental activism given by Chernobyl testifies to a healthy and permanent "greening" of East European politics.

Leadership Change and Gorbachev's Policies

The final event of the 1980s so far to affect environmental politics in Eastern Europe is the accession of Mikhail Gorbachev to the leadership of the Soviet Union. *Perestroika* calls for energy efficiency and an end to Stalinist economic practice. *Glasnost* (openness) has encouraged the proliferation of independent environmental groups and greater openness in the dissemination of environmental information. Beyond Soviet borders, these twin developments have laid bare an array of problems that go to the heart of the present political system in Eastern Europe. Free information and economic reform are needed before environmental problems can be solved, yet both render the current leadership politically vulnerable. It is not surprising therefore that Gorbachev's call for reform has elicited a greater response in the younger generation. In an op-ed article on the Prague Spring in the *International Herald Tribune*, Jiri Pehe argued that the reform movement in Czechoslovakia was spearheaded by a younger generation of middle-class professionals and intellectuals born during and after World War II, less ideological and more flexible than their seniors. The core of Gorbachev's supporters in the Soviet Union being drawn from the same social stratum and age group, *perestroika* thus poses a generational as well as a programmatic challenge to the USSR's East European allies.[68] A younger and more educated middle class is enthusiastically welcoming reforms that, even in Hungary, are being only reluctantly endorsed by the senior party leadership.

The *perestroika* generation gap may be found in the institutes and research centers of Eastern Europe. Young researchers daily confront the reality of hierarchy in the form of senior administrators and established scholars who have made their adjustment to the system. The environmental area is no exception to this rule. Environmental education is required throughout Eastern Europe. It is supplemented by the numerous programs of the mass environmental organizations directed at teaching young people the importance of environmental protection and the dangers of disturbing the ecological balance. Young people assist at monitoring stations, take air and water samples, identify flora and fauna in nature preserves, apprehend violators of park and preserve regulations and turn them over to the authorities, and take part in wildlife management. Participation in such activities tends to foster bonds of common interest and friendship, the basis of the "invisible colleges" of the adult years. In all the countries of Eastern Europe young intellectuals form the nucleus of the independent environmental groups. In Czechoslovakia, I found that those who had been working in the environmental area since the 1950s tended to be optimistic about the government's approach to reducing pollution. By contrast, younger scholars appeared pessimistic and discouraged. "To solve environmental problems," they said, "you need a workable program and free speech. Our countries have neither."

In Yugoslavia, where university graduates sometimes spend years waiting for placement in an institute, the generation gap may be even more serious. Young people there cannot understand why individuals with no specialized education who fought a successful guerrilla war forty years ago should still be at the head of major institutions. The middle generation that is slowly taking the partisans' place is perceived as ill-equipped to exercise leadership. There is considerable feeling that unless the educated and vital younger generation replaces the veterans of the war of national liberation, there will be no change in Yugoslavia. In this context, the demonstrations in Ljubljana sponsored by the Slovenian Communist Youth League take on more significance. As one demonstrator told me, a major reason for his participation was to oppose the established regime. Although environmental or nuclear issues may be exploited as a means to vent frustration, there is little doubt that generational frustration has aided and abetted the public airing of ecological problems.

CONCLUSIONS

It is risky to attempt to generalize about countries as diverse as those of Eastern Europe. But the evidence does suggest a common thread of development in the environmental politics of these countries. The increased visibility and influence of the noneconomic actors in environmental management in Eastern Europe in the 1980s mark a watershed in environmental politics in the area. The development of an independent environmental lobby doubtless has been facilitated by developments in Moscow, although this process started before the leadership change in the Soviet Union. The most important impact of Gorbachev's advent has been to intensify demands for generational change. Given the very negative ecological predictions for Eastern Europe well into the next century,[69] the environmental lobby will probably gather more and more popular support, consolidate its organization, and further internationalize its orientation. As it gains in strength, it may be expected to win over substantial numbers of the new generation of party leaders and industrial managers who face economic catastrophe in the approaching environmental crisis.

NOTES

1. See Bogdan Poplawski, "Environmental Protection and Development in Poland in the 1970s," *Gospodarka planowa*, no. 3 (March 1980): 131–39.
2. V. Mezricky and Collective, *Životní prostředí: véc veřejné i soukromá* (Prague: Prace, 1986), pp. 259–62.
3. See Barbara Jancar, *Environmental Management in the Soviet Union and Yugoslavia: Structure and Regulation in Communist One-Party States* (Durham, N.C.: Duke University Press, 1987), app. 2, pp. 352–53.
4. For a complete summary of environmental legislation in Eastern Europe, see also György Enyedi, August J. Gijswijt, and Barbara Rhode, eds., *Environmental Policies in East and West* (London: Taylor Graham, 1988) (hereafter cited as *Environmental Policies*).
5. Alec Nove, *The Soviet Economic System*, 2nd ed. (London: George Allen & Unwin, 1982), p. 321.
6. *Radio Free Europe (RFE) Research*, "East European Leadership List" 13, no. 2, pt. 3 (15 January 1988).
7. "Procedure in Authorizing Publication," *Věstník Ministerstva zdravotnictví*, no. 3 (1982): 23–24.
8. For an analysis of the republican component in Yugoslavia's environmental politics, see Barbara Jancar, "Environmental Protection: The Tragedy of the Republics," in *Yugoslavia in the 1980s*, Pedro Ramet, ed. (Boulder and London: Westview Press, 1985), pp. 224–46.
9. *Vjestnik*, 9 November 1980, p. 5.
10. Lest readers doubt the stubborn persistence of rigid dogmatism among the older generation of East European Communists, they are referred to the recently published and fascinating study of the political attitudes of the Communist leaders of postwar Poland: Teresa Torańska, *Them*, trans. Agnieszka Kołakowska (New York: Harper & Row, 1987).
11. Compare Aleksandra Jasińska and Renata Siemieńska, "Kadra kierownicza o problemach społeczności lokalnych i własnej działalności," in *Władza lokalna u progu kryzysu*, Jerzy J. Wiatr, ed. (Warsaw: Warsaw University, Sociological Institute, 1983), pp. 254–55; and Renata Siemieńska, "Wartości, aspiracje i oczekiwania społeczeństwa polskiego a perspektywy zaspokajania potrzeb," in *Zaspokajanie potrzeb w warunkach kryzysu*, Jerzy J. Wiatr, ed. (Publication no. 17; Warsaw: Warsaw University, Sociological Institute, 1986), p. 525. The entire chapter (pp. 516–53) is worth reading for its information on social attitudes in Poland today. The volume was published in only 100 copies without censorship.
12. For the story of the closing of the plant, see "Poland: the Most Polluted Country in the World?" *New Scientist* 92, no. 1276 (22 October 1981), 248–50; and Vera Rich, "Smelter Shuts Down," *Nature* 258 (15 January 1981): 112.
13. During the 1976–80 plan in Poland, 55 billion zlotys were earmarked for environmental protection, of which only 38.8 (75 percent) billion were actually used. By contrast, 72 percent of the planned 15 billion zlotys were used during the 1966–70 plan and 72 percent of the planned 34 billion zlotys allotted in the 1971–85 plan. (Warsaw, *Perspektywy*, no. 4 (25 January 1980): 11. Evidence of the stubborn persistence of rigid dogmatism among the older generation of East European Communists is provided by Teresa Torańska, *Them*.
14. See Jancar, *Environmental Management*, chap. 5.
15. Guy Benveniste, *The Politics of Expertise*, 2nd ed. (San Francisco: Boyd and Fraser, 1977), pp. 4–5.
16. Sharon Wolchik, "The Scientific-Technological Revolution and the Role of Specialist Elites in Policy-Making in Czechoslovakia," in *Foreign and Domestic Politics in Eastern Europe in the 1980s: Trends and Prospects*, Michael J. Sodaro and Sharon L. Wolchik, eds. (London: Macmillan & Co., 1983), p. 112.

17. See Robert Rich, ed., "Symposium on the Production and Application of Knowledge," *American Behavorial Scientist* 22, no. 3 (January–February 1979): 420–34.
18. H. Gordon Skilling, *Interest Groups in Soviet Society* (Princeton: Princeton University Press, 1971); Donald R. Kelley, "Interest Groups in the USSR: The Impact of Political Sensitivity on Group Influence," *The Journal of Politics* 34, no. 3 (August 1972): 860–88; Peter H. Solomon, *Soviet Criminologists and Criminal Policy: Specialists in Policy-Making* (New York: Columbia University Press, 1978); and John Lowenhardt, *Decision Making in Soviet Politics* (New York: St. Martin's Press, 1981).
19. Thane Gustafson, *Reform in Soviet Politics: Lessons of Recent Policies on Land Water* (Cambridge University Press, 1981), chaps. 3 and 4.
20. For a short description of the work of the Polish League for the Preservation of Nature expressed by Edward J. Kormondy, see "Environmental Protection in Hungary and Poland," *Environment* 22, no. 10: 37. The efficacy of the league was questioned by František Hron, Chairman, Czech Union of the Department of Nature, in an interview in May 1987.
21. Interview with Otto Inzsel, Chair, Environmental Committee of the Patriotic People's Front, August 1986. See also Kormondy, "Environmental Protection," 34–35.
22. CSOP, "Stanovy Českého svazu ochrancu přírody" (Prague: Ústřední výbor Ceskeho svazu ochrancu přírody, 1985), para. 1.2, p. 3.
23. Author's visit to the State Farm at Vysoké n/Jiz, May 1987.
24. *Conception and Requirement System for Environmental Protection in Hungary* (Budapest: OMKDK, 1984).
25. See the editorial describing the program in *Rude Pravo*, 19 November 1985; and CSSR law no. 65/1986 Sb.
26. *Rzeczpospolita*, 24 June 1986, as cited in "Independent Groups Warn of Potential Ecological Catastrophe," *RFE Research*, Polish SR/13, 29 August 1986, p. 39.
27. Enyedi et al., *Environmental Policies*, pp. 57–59.
28. The first meeting of experts from countries responding to Zhivkov's proposals of a Balkan ecological pact (Albania, Bulgaria, Greece, Romania, and Yugoslavia) took place on 16 and 17 December 1987 in Sofia. (Radio Free Europe Research, *RAD Background Report*, BR/5, 13, no. 3 [22 January 1988], p. 7.) For the Danube convention proposal, see *Rabotnichesko delo*, 21 February 1988.
29. Radio Budapest, 28 May 1986.
30. The agreement was reached in 1984.
31. For the Czechoslovak side of the controversy, see K. Noha, "Hainburg or Wolfsthal-Bratislava?" *Rude Pravo*, 22 January 1985, p. 5.
32. For a description of the issues involved in cleaning up the Elbe River, see "Can Two Germanies Clean Up One River?" *New Scientist* 92, no. 1280 (19 November 1981): 484. See also "The Unshared Dirt," *Die Zeit*, 6 March 1987. For more on German-German cooperation, see the chapter by Helmut Schreiber in this volume.
33. *Tygodnik Mazowsze* (Warsaw), no. 170, 15 May 1986.
34. By the end of 1986, the Hungarian government had paid out about 408 million forints in compensation, covering about half the actual losses (*Heti vilaggazdaság*, 13 December 1986).
35. Statement by Mieczysław Sowiński, Chairman of Poland's Atomic Energy Agency, PAP, 13 June 1986.
36. RFE Correspondent's report (Vienna), 12 February 1987.
37. Jancar, *Environmental Management*, chap. 6.
38. Mezricky, *Životní prostředí*.
39. *Rzeczpospolita*, 19 January 1987.
40. The exhibition was reported in *Dunavska Pravda* (Ruse), 6 December 1987. For a Western account of the exhibit, see *RFE Research*, Bulgarian SR/2 (11 February 1988), part 6, "Air Pollution in Ruse Reaches Appalling Levels," pp. 29–31.
41. Petr Privetivy, "Utopime srdce Krovoklatska?" *Naší přírodou*, December 1986, pp. 275, 282.

42. *Bratislava nahlas* was released 4 June 1987 and copies reached the West in the fall. Official criticism of the report may be found in *Pravda* (Bratislava), 4 February 1988, p. 3.
43. "The Tenth Anniversary of Charter 77," *RFE Research*, Czechoslovak SR/1, 17 January 1987, p. 3.
44. Charter 77, document no. 13, 1986.
45. Charter 77, document no. 1, 1987.
46. Charter 77, document no. 33, 30 April 1987.
47. For a sample of Polish underground reporting of Chernobyl, see *Extracts from Polish Underground Publications*, Nika Krzeczunowicz and Anna Pomian, comps. and eds., *RFE Research*, Polish Underground Extracts/8, June 1986.
48. For discussion of the concept of "loyal opposition," see Barbara Jancar, "The Case for a Loyal Opposition under Communism: Czechoslovakia and Yugoslavia," *Orbis* (summer 1968): 415–40.
49. "Campaigning for Environmental Protection," *RFE Research*, Polish SR/5, 15 May 1987, p. 30.
50. Magdalena Mil'tsazh, "Conservation Club," *Sotsialisticheskaya industriya*, 29 August 1979, p. 3.
51. For published references to the early days of the Polish Ecological Club, see "Poland, the Most Polluted Country," 246; and Rich, "Smelter Shuts Down," 112.
52. Much of this information comes from author's interviews in Poland. Published confirmation may be found in the RFE summary of the development of independent environmental groups entitled "Independent Groups Warn of Potential Ecological Catastrophe," *RFE Research* SR/13, 29 August 1986.
53. For first-hand Western accounts of the Warsaw Freedom and Peace conference, see Polly Duncan, "A New Generation of Opposition," *Sojourners: An Independent Christian Monthly* 16, no. 9 (October 1987): 14–25; and Janet Fleishman, "Beyond the Blocs? Peace and Freedom Hosts International Seminar in Warsaw," *Across Frontiers* 3, no. 4 (summer–fall 1987): 7–13.
54. As reported in *Die Welt*, 23 November 1984; and *Die Weltwoche*, 29 August 1985.
55. The appeal is summarized in Edith Markos, "Protests Continue against Danube Hydroelectric Project," *RFE Research*, Hungary SR/11, 14 October 1985, pp. 23–7.
56. For an overall account of the environmental movement in Hungary, see "Special Report on Environmental Politics in Hungary," *Across Frontiers* 3, no. 4 (summer–fall 1987): 7–13.
57. See Charter 77, document no. 33, 1987; and *Novo Slovo*, no. 7, 1988, pp. 4–5.
58. Associated Press (Vienna), 5 December 1984; *Die Presse*, 28–29 June 1986.
59. For a report on the activities of the Czech Society of Defenders of Nature in the first two-and-a-half years of its existence, see Peter Přívětivý, "Mezi dvéma polo-časy," *Naší přírodou*, December 1987, p. 97.
60. A list of the organizations in the Czech National Front authorized to engage in environmental protection activities may be found in Rada pro životní prostředí při vládě CSR, *Příručka pro ochránce přírody*, 2nd ed., Informational Publication 14/1984 (Prague: Státní zemědělské nakladatelství, 1984), p. 28.
61. Before that time, I had given two press interviews on environmental issues of mutual interest to Czechoslovakia and the United States. The first was published in the October 1987 edition of the Czech Society of Defenders of Nature's monthly *Nasi prirodou*, the second in the party weekly *Tvorba (Creation)* in February 1988.
62. *Rude Pravo*, 28 May 1986.
63. For a record of the protest and its impact, see *Pruboj* (20 September 1987) supplement (26 September 1987), p. 9; and 3 October 1987, p. 3 and supplement.
64. *Literaturen Front* (Literary Front), no. 8 (18 February 1988).
65. See Jancar, *Environmental Management*, chaps. 5–7.
66. See Jancar, *Environmental Management*, "Postscript Chernobyl."

67. For further discussion of environmental groups in Eastern Europe, see *From Below: Independent Peace and Environmental Movements in Eastern Europe and the USSR* (New York: U.S. Helsinki Watch Committee), October 1987.

68. *International Herald Tribune*, 24 May 1987.

69. The Deputy Head of Krkonoše National Park in Czechoslovakia indicated he did not anticipate any major change in ambient air quality over the park until at least the year 2020. Rather than try to save a monoculture that could probably not be saved, he said, the major strategy was to replant with trees that could bear the high level of enviromental stress.

2

ENERGY AND THE ENVIRONMENT IN EASTERN EUROPE

John M. Kramer

Three preeminent factors determine the nexus between energy policies and the environment in Eastern Europe.[1] First, economic planners in the region traditionally have emphasized "supply-side" policies to expand the amount, rather than "demand-side" policies to limit the consumption, of energy. Consequently, the East European economies have developed, in the words of Hungarian economists, an "insatiable appetite" for fuels and power that engenders among them a "persisting propensity to overconsume" these resources.[2] All states in the region continue to devote between 40 and 45 percent of their total investments in industry to increase the indigenous production of fuels and power.[3] The quality of the environment suffers considerably from these policies. The more energy an economy consumes (all other things being equal), the more wastes it creates, and so the more it pollutes the environment. Further, the enormous output of fiscal resources that these states devote to increasing their supply of energy severely constrains their capacity to pursue other investment opportunities, including opportunities to protect the environment.[4]

Second, policies designed to increase the indigenous production of energy have generally meant increased reliance on highly polluting brown coal and lignite, because reserves of oil, natural gas, and hard coal in the region are limited.[5] The consequences are manifest: so-called lunar landscapes are created by the strip-mining of brown coal and lignite, and emissions of sulfur dioxide from brown-coal combustion have intensively polluted the air, land, and water in many areas. Coal has contributed more to the degradation of the environment in Eastern Europe than all other forms of primary energy combined. Hence the future status of coal in primary energy consumption will exercise a decisive influence on the quality of the environment in the region.[6]

Third, Eastern Europe must rely primarily on importation from the USSR for oil and natural gas, the use of which typically entails less damage to the environment. Consumption of these fuels in Eastern Europe increased rapidly in the 1970s with attendant benefits for the quality of the environment. Increased use of natural gas has continued in the 1980s and will accelerate through the remainder of this century, as major pipeline projects are completed to bring Soviet gas from Western Siberia to markets in Eastern and Western Europe. The future of Soviet oil exports to Eastern Europe is far less bright. This circumstance could benefit the environment if it encourages intensified energy conservation efforts and limits the consumption of oil by private motorists. East European states are more likely, however, simply to increase their production of highly polluting brown coal and lignite to compensate for reduced deliveries of Soviet oil.

In sum, if the quality of the environment in Eastern Europe is to improve substantially, ways must be found to reduce consumption of energy in general and of highly polluting solid fuels in particular. To assess the likelihood that these conditions will be realized, we must first understand the factors promoting excessive consumption of energy and then examine the present—and likely the future—status of the principal forms of primary energy resources consumed in this region.

THE CONSUMPTION OF ENERGY

Overall, East European and Soviet data indicate that the states of communist Europe on average consume 30 to 50 percent more energy than do industrialized capitalist countries to produce similar units of national income.[7] Table 2.1 illustrates that East European states, even with far smaller national economies and stocks of motor vehicles, have levels of per capita energy consumption that equal, and often exceed, those of many Western economies.

High energy consumption results partly from the disproportionate share of solid fuels of low caloric content in most energy balances in the region. These fuels incur much greater losses of energy in extraction, transportation, and conversion into usable forms of power than do other forms of primary energy. Furthermore, inefficient and obsolete power equipment abounds in the region. For example, in Poland, where thousands of industrial furnaces operate at below 40-

Table 2.1
ENERGY CONSUMPTION FOR SELECTED COUNTRIES
(Barrels of Oil Equivalent per Capita), 1984

Country	Barrels per Capita
GDR[a]	43
Czechoslovakia	36
FRG[b]	32
Bulgaria	29
France	26
Poland	25
UK[c]	25
Romania	23
Japan	23
Hungary	21

SOURCE: Computed from data in Directorate of Intelligence, *Handbook of Economic Statistics, 1986* (Washington, D.C.: Government Printing Office, 1986), CPAS86-10002, tables 2, 3, pp. 24–25.
[a]GDR = German Democratic Republic.
[b]FRG = Federal Republic of Germany
[c]UK = United Kingdom.

percent efficiency, it would take between sixty and eighty years at current levels of production to modernize the national stock of industrial and residential power equipment.[8]

Excessive consumption of energy is inextricably related to the Stalinist model of economic development and organization—the so-called command economy, which the USSR imposed uniformly throughout the region after 1948. The model viewed factor inputs, including fuels and energy, as inexhaustible and saw high levels of fuels and power consumption as indexes of modernization. The task of economic planners was to ensure that society's resources were mobilized in an all-out assault to achieve ever-higher economic targets. Consequently, the Stalinist model sought economic development primarily through "extensive" growth (quantitative increases in factor inputs) rather than through "intensive" growth (qualitative increases in factor productivity). Although the "pure" Stalinist model has been modified by the post-Stalin regimes in Eastern Europe, many of its features persist and foster excessive consumption of energy.

The imperative to fulfill plan targets is among the most important of these features.[9] Production personnel, as rational economic actors, predictably try to maximize economic gains and minimize losses: because fulfillment of the plan is the road to bonuses, promotions, and other perquisites, enterprise managers naturally subordinate all other objectives (including energy conservation) to this end. In fact, producers possess economic incentives to maximize, and not minimize, their expenditure of energy and other inputs, so as to inflate the gross output or other value indicators typically employed to measure fulfillment of the plan.[10]

The pricing system is also a legacy of the Stalinist model. Prices are established administratively and reflect the preferences of political elites rather than the dictates of supply and demand. Prices of energy resources traditionally have been set so low that it is economically difficult to justify conservation efforts.[11] This pricing policy reflects several factors: the ideological commitment to the Marxist theory of value, which considers natural resources "free goods"; a desire by planners to encourage the production of commodities with a high energy content; and a concern to avoid unpopular price increases.

In lieu of economic incentives to conserve, officials typically establish administratively determined norms for consumption, which industrial consumers are obligated to observe. The defects in this approach are manifest. First, the norms frequently are inflated because producers themselves—having obvious incentives to overstate their requirements for energy—provide the data upon which they are based.[12] Second, penalties for noncompliance are minimal and, in any case, rarely enforced.[13] Third, technologically substantiated norms for energy consumption do not exist for many products and production processes, partly because production personnel have little interest in devising them and partly because of shortages of requisite measuring and control equipment to implement them.[14]

Finally, the Stalinist model has promoted excessive energy consumption through its almost obsessive preoccupation with heavy industrial development (chemicals, steel, and other highly energy-intensive industries) and its disregard for less energy-intensive sectors (for example, the service sector). Escalating world market prices for energy after 1973 did little to alter these conditions: it was precisely in the mid-1970s that many of the most oil-intensive technologies and equipment (petrochemicals and refinery capacities, for example) were introduced into Eastern Europe, and national economic plans for 1981–85 typically projected that energy-intensive industrial sectors would grow more rapidly than the economy as a whole.[15]

In response to the well-known developments in the price and supply of energy on the international market in the late 1970s, all East European regimes (albeit with varying degrees of enthusiasm and com-

Table 2.2
AVERAGE ANNUAL CHANGE IN UNIT ENERGY REQUIREMENTS
per Unit of National Income (Percent), 1981–83

Country	Percent
United States	−3.1
Japan	−4.7
EEC[a]	−2.7
CMEA[b]	−1.7
Bulgaria	−1.6
Czechoslovakia	−0.6
Hungary	−2.5
GDR[c]	−2.9
Poland	+0.9
Romania[c]	−3.8

SOURCES: *Magyarorzag*, 7 July 1985, in JPRS-EEI, 20 September 1985, pp. 28–29; *Figyelo*, 27 June 1985, in JPRS-EEI, 9 August 1985, pp. 22, 24.
[a]EEC = European Economic Community.
[b]CMEA = Council for Mutual Assistance.
[c]This table should be treated with extreme caution because of the notorious unreliability of official Romanian data overall, and of data relating to national income in particular.

mitment) have now initiated programs to reduce their consumption of energy.[16] The substance of these programs is beyond the purview of the present study. Suffice it to say that, in the words of East European sources, in general these programs have to date encompassed "only the most easily mobilizable reserves." A "better housekeeping approach" has been the "likely" source of whatever energy has been conserved, and the response of enterprises and households to the imperatives of conservation has been "strikingly inadequate."[17] With the exception of the GDR, stagnant or declining rates of economic growth—and not conservation measures—appear to be the primary factor limiting demand for energy in several East European economies in the 1980s.[18] Overall, in the assessment of a Hungarian source, only the GDR and Hungary can at present show "internationally competitive trends" in the conservation of energy.[19] (See table 2.2.)

PRIMARY ENERGY RESOURCES AND THE ENVIRONMENT

Coal

At least through the mid-1970s, Eastern Europe appeared to be following the path of Western Europe and the United States, where economic development has led to diminished reliance on coal and greater reliance on liquid fuels in primary energy consumption. In 1960 coal constituted 84 percent of the primary energy consumed in Eastern Europe, compared to 57 percent in 1977.[20] Two related factors prompted the reduced reliance on coal. First, mining conditions worsened as higher quality and easily exploitable deposits were depleted, increasing the expense of extraction and decreasing the cost-effectiveness of coal relative to other forms of primary energy.[21] Second, Soviet oil and gas seemed to offer readily available and relatively inexpensive alternatives. The import of Soviet oil began on a large scale in the late 1960s, and of natural gas in the mid-1970s.

Reliance on coal revived in the wake of the twin oil shocks on the international market in 1973 and (especially) 1979. Economic planners looked more favorably upon coal as the price of oil on the capitalist market became prohibitive and Soviet prices also increased substantially.[22] Years of neglecting the coal industries in the region could not be reversed rapidly, however. Consequently, plan targets and results for coal production in 1976–80 were at best modest for all countries in the region with the partial exception of Romania. Overall production during this period increased by approximately 7 percent, and only in Romania did output of coal in 1980 exceed the level attained in 1975 by more than 8 percent.[23] Bulgaria, the GDR, and Romania all underfulfilled their plan targets for coal to varying degrees. Poland may well have done so too; Polish officials now acknowledge that coal output figures were grossly inflated in these years by more than 20 percent over actual production.[24]

The enormous increase in world energy prices in 1979–81 provided the real imperative to increase reliance on coal. Accordingly, all national economic plans for 1981–85 projected increases in coal production, although the rates of increase varied considerably among them. Romania established by far the most ambitious (and unrealistic) of these targets, seeking increases of 143 to 148 percent in the rate of coal production in order to raise the share of coal in the generation of electricity from 27 to 50 percent between 1980 and 1985.[25] At the other extreme, Hungary sought to increase production in 1985 by just 2 percent over the 1980 level, although realization of even this modest target would require allocating to coal mining 7.5 percent of total industrial investments in the plan period, as compared to 3.5 and 4.7 percent in the preceding two plan periods.[26]

Despite planned expansion, Bulgaria, Czechoslovakia, Hungary, and Poland each produced less hard coal in 1985 than in 1980 and less brown coal and lignite in 1985 than in 1984.[27] Nonetheless, overall production of coal in Eastern Europe was almost 13 percent greater in 1985 than in 1980, and the respective figures for brown coal and lignite were almost 20 percent greater. Further, production of brown coal and lignite in the GDR grew from 258 million tons in 1980 to 312 million tons in 1985—the latter figure exceeding the plan target by 22 million tons. Romania also experienced a substantial increase in production from 27.1 million tons in 1980 to 37.6 million tons in 1985, although output in 1985 was far below the utterly unrealistic plan target of 87 million tons.

Sharp differences appear in the status of coal in the national economic plans for 1986–90. All states except Czechoslovakia and Hungary plan overall increases (comprised almost exclusively of brown coal and lignite) in the production of coal.[28] Romania again distinguishes itself with the most ambitious and unrealistic of these targets: an overall increase in production by 1990 to between 150 and 165 percent of the level of output attained in 1985. Bulgaria and Poland also plan substantial increases in production that would raise the level of output of brown coal and lignite in 1990 by nearly 40 percent compared to 1985. Hard-coal mining in Poland, on the other hand, has encountered (in the words of the official Polish news agency) "a ceiling of technology and organization" that likely will prevent any increases in production at least through 1990 and perhaps through the end of the century.[29] The GDR plans a more modest increase in production—from 312 million tons in 1985 to between 330 and 335 million tons by 1990—but this target must be assessed in the context of the huge increases in production that have occurred since the mid-1970s (including 1985, when the plan target for output was overfulfilled substantially).

The prospects for coal in Czechoslovakia are particularly interesting from the standpoint of the environment. Czechoslovakia now plans to reduce brown-coal and lignite production in 1990 to 90 percent, and in 2000 to 80 percent, of the 1985 level.[30] Czechoslovak sources explicitly identify a desire to limit further damage to the environment as one among several factors—albeit probably not the most important one—prompting the projected decline in output.[31] To compensate for reduced coal supplies, planners project increased nuclear power capacities by 1990 and foresee an average annual decline in the energy intensity of the national economy of 2.1 percent in 1986–90. Given the past performance of Czechoslovakia in meeting its targets for nuclear power and making the economy more energy-efficient, one must express reservations that the desired reduction in coal output will be realized.

Continued reliance on coal in Eastern Europe constitutes a preeminent obstacle to any substantial improvement in the quality of the environment. In all countries of the region, except Hungary and perhaps Czechoslovakia, coal will either retain or increase its share in primary energy consumption through the year 2000. Even more threatening to the environment is the increasing share of brown coal and lignite of low caloric value and high sulfur content in the coal being consumed. The Polish case illustrates this dilemma. As noted, Poland wishes to expand extraction of brown coal and lignite partly to compensate for constraints on production of hard coal, and partly to allow higher exports of hard coal, its most important earner of convertible currency. The brown coal and lignite now being exploited have a sulfur content as high as 7 percent, versus 1 to 1.5 percent in Silesian hard coal, and an energy yield that is between one-half and two-thirds less than its hard-coal counterpart.[32] By one estimate, expanded use of low-quality coal could increase emissions of sulfur dioxide in Poland by as much as 80 percent in 1983–90. Some believe the opening of new mines and expansion of existing ones is likely to result in "total environmental degradation in the affected areas."[33]

Oil

Oil has proved a mixed blessing to the environment in Eastern Europe. The environment clearly has benefited when oil has replaced coal in primary energy consumption, a process particularly pronounced in the 1970s. Yet motor vehicles—those seemingly ubiquitous polluters in modern societies—stand as a powerful symbol of the threat that oil represents to the environment. Although motorization is far less advanced in Eastern Europe than in Western Europe and the United States, it is a primary polluter in urban centers. In Budapest, for example, motor vehicles account for 60 to 70 percent of the emissions of nitrogen dioxide and carbon monoxide into the atmosphere; and in Warsaw, where exhaust fumes comprise more than 70 percent of the toxic gases in the atmosphere, about 60 percent of the city risks levels of carbon monoxide more than double the permissible norm by 1990.[34]

Importation from the USSR has been the principal source of oil for all countries in the region except Romania, which purchases most of its oil on the world market and also meets a significant (albeit declining) share of its requirements through domestic production.[35] Soviet oil began flowing to Eastern Europe in large quantities with the completion of the *Druzhba* (Friendship) oil pipeline in the early

1960s. By 1967, the Druzhba network was carrying approximately 18 million tons of crude annually from the USSR to Eastern Europe, via a northern branch to Poland and East Germany and via a southern branch to Czechoslovakia and Hungary. These exports expanded so dramatically in the following decade that during the 1976–80 plan period the USSR shipped to Eastern Europe more than 370 million tons of crude oil and oil products.[36] Consequently, by the late 1970s oil comprised approximately 25 percent of the primary energy consumed in the region as compared to 8 percent in 1960.[37] Entering the 1980s, East European leaders had reason to believe that Soviet oil deliveries during the 1981–85 plan period would continue at previous or higher levels.[38] Suddenly in 1982, however, the USSR announced a reduction of about 10 percent in its planned oil deliveries to Eastern Europe in 1982–85.[39] This action was apparently prompted by important economic considerations. Soviet commentators have expressed doubts that domestic oil production will continue to be sufficient simultaneously to satisfy internal demand, meet national requirements for hard currency through exports to the world market, and serve as the primary source of supply for Eastern Europe.[40] The precipitous decline in world oil prices since 1981 sharpened the tensions inherent in fulfilling these several commitments. Exports of energy, particularly oil, to the world market are the principal source of hard currency for the USSR, which in turn is used to finance such high-priority projects as the acquisition of Western technology and machinery to modernize Soviet industry, and of grain to feed the Soviet population. Hence the USSR responded to declining oil prices, in the words of one Western analyst, "by frantically trying to remain in the market, even at the expense of its allies." In 1982 the USSR increased by 32 percent, and in 1983 by another 15 percent, the volume of its petroleum exports to advanced capitalist countries while simultaneously reducing by 10 percent its petroleum exports to Eastern Europe.[41] East European sources explicitly link the reduction in Soviet oil exports to Eastern Europe with the increase in exports to the capitalist market.[42]

The USSR also responded to declining hard-currency revenues by pressing the East Europeans to pay more for Soviet energy. Prices for oil have been among the most controversial issues in intra-CMEA trade since the initial escalation in world energy prices after October 1973. At that time, CMEA still employed the "Bucharest formula," whereby prices for oil (and other commodities) were fixed in 1971–75 and reflected the average world market prices for this commodity in 1966–70. By March 1974, with the official CMEA price for oil already 80 percent below the comparable world price, the USSR demanded an immediate revision of the formula. The new system allowed Moscow a special price increase for 1975 and provided for subsequent annual adjustments in prices based on average world market prices for individual commodities in the preceding five-year period.[43]

The new price system did result in a substantial increase in the CMEA price for oil, although this price consistently has lagged behind the comparable world market price. One Western analyst estimates that whereas in 1978 the CMEA price for oil was 53 percent of the world price, by 1984 the respective figure was 41 percent.[44] Eastern Europe actually benefited even more than these data suggest, because it traditionally paid for most Soviet energy with "soft" goods (i.e., goods not competitive on the world market) that often had artificially high prices attached to them.[45]

The price and availability of Soviet fuels and energy have become prominent topics in the 1980s at meetings of the CMEA leaders. Thus, at the 1983 Session of the Council (CMEA's principal political organ), Premier Nikolai Tikhonov of the USSR made clear his country's desire for changes in CMEA's energy trade. He asserted that "mutual advantage" must prevail in this trade and that Eastern Europe must take "urgent measures" to balance trade with the USSR by supplying it with "goods intended for export" (i.e., goods competitive on the world market, or "hard" goods). "Understandably," he continued, the willingness of the USSR to meet Eastern Europe's demand for energy "depends considerably on the extent to which the CMEA countries provide goods that are needed by the national economy of the USSR."[46]

It is not surprising that the East Europeans, while acknowledging that market conditions justify higher prices for Soviet energy, have sought to minimize the extent of the price increases and ensure that the USSR honors its export commitments to them. The most cogent of their arguments is that precipitous price increases for, and/or failure to deliver, contracted amounts of vital raw materials could engender serious economic dislocations that may spark political unrest and threaten the already fragile political stability of the socialist regimes.[47] Premier Lubomir Strougal of Czechoslovakia noted that "the economies of our individual states react very sensitively to shortcomings in the development of mutual trade," and he pointedly remarked that any "difficulties and complications in this sphere would have a chain reaction."[48] Hungarian sources also point out that the USSR itself is guilty of paying below-world-market prices for the hard goods (e.g., foodstuffs) it receives in payment for oil. "We should not be made to suffer unnecessary losses," argued the Hungarian delegate to the 1985 CMEA Session of the Council in Warsaw, "in connection with ever growing energy prices."[49] Nonetheless, prospects are minimal for any expansion of Soviet oil exports to Eastern Europe in 1986–90. Publicly, the USSR has pledged that it will maintain its annual deliveries of oil during this period at the 1985 levels.[50] Cynics may note that the USSR made a similar pledge to honor its export commitments shortly before enacting the substantial reduction in oil exports to the region in 1982.

Eastern Europe has responded to higher prices for Soviet oil in ways that both harm and benefit the environment. Regimes have sought whenever feasible to substitute solid fuels for increasingly costly liquid fuels, despite the implied negative consequences for the environment; at the same time, however, they have also tried to conserve oil by enacting substantial increases in the price of gasoline to limit private motorization. Thus in Hungary, although the national stock of automobiles expanded steadily in 1981–84, gasoline consumption declined by 7 percent in the face of price increases of 26 percent.[51] The region's environment has also benefited from the increasing substitution of natural gas, primarily imported from the USSR, for oil in primary energy consumption.

Natural Gas

Natural gas, as one of its proponents points out, is an "environmentally friendly kind of fuel" with limited potential for pollution.[52] Increased consumption of natural gas in the 1970s reportedly contributed to a marked improvement in the environment in parts of Eastern Europe.[53] By 1977, natural gas constituted 16 percent of the primary energy produced in the region (versus just 6 percent in 1960). The use of natural gas was concentrated in Hungary and Romania, however, where it represented 23 percent and 51 percent of primary energy production respectively (versus less than 9 percent in the other East European countries). Domestic production of natural gas covered most of the increased use in all East European states through the 1970s—except Bulgaria and Czechoslovakia, which possess minimal reserves. This circumstance has changed in the 1980s, however, as indigenous production has either stagnated or declined.[54]

Any substantial increase in consumption of natural gas in the region now depends on expanded imports from the USSR. The potential for such an expansion is bright: the production of natural gas in the USSR, in sharp contrast to that of oil, has been increasing dramatically since the 1970s, and known Soviet reserves—more than one-third of the proved natural gas reserves in the world—are ample to meet projected production targets through 1990.[55]

The USSR began exporting small quantities of natural gas via border-crossing pipelines to Poland in the late 1950s and to Czechoslovakia in 1967, but it was only with the completion of the *Bratstvo* (Brotherhood) natural gas pipeline network linking gas fields in the Soviet Ukraine to Eastern and Western Europe that gas began to be shipped in significant volumes. Overall, natural gas grew from just 3 percent to more than 16 percent of total Soviet energy exports to Eastern Europe between 1970 and 1980. By 1980, Soviet imports accounted for approximately one-third of the natural gas consumed in the region, but only 6 percent of the primary energy.[56]

The East European states participate actively in "joint projects" with the USSR to exploit Soviet gas deposits and to construct pipelines and transport facilities. Many of these pipeline projects are designed to export Soviet natural gas not only to Eastern Europe but to Western Europe as well. Typically, the East European participants in these so-called joint projects provide low-interest investment credits, manpower, equipment (some of it imported from capitalist countries in exchange for hard currency), and technical assistance; they receive in return an agreed-upon volume of natural gas for a fixed period.[57]

The Orenburg natural gas pipeline network, in which all the East European members of CMEA participated, represents the most important of these joint projects completed to date. The project involved the construction of a 1,733-mile pipeline stretching from the Orenburg gas fields in western Siberia to existing pipelines at the Soviet Union's western border. The contribution of the East Europeans to the project included providing financial credits, machinery, and more than 20,000 construction workers. When completed in 1979, the pipeline had a capacity of 28 billion cubic meters per year, approximately 55 percent of which the East Europeans will receive annually for twenty years as repayment for their participation. The overall amount of natural gas to be received is equal to approximately one-half the total proved reserves of natural gas in Eastern Europe.[58]

Construction of the Siberian gas export line to Western Europe also involved three East European states: Hungary, Yugoslavia, and particularly Czechoslovakia. In 1983 Czechoslovakia began constructing yet another pipeline across its territory, to be completed in 1988, to bring Soviet gas to West Germany. The Soviet Union is to reimburse Czechoslovakia with natural gas. According to one Western estimate, Czechoslovakia could receive annually in natural gas the calorific equivalent of approximately seven million tons of high-quality brown coal.[59]

The latest of the joint projects involves the exploitation of the giant gas field at Yamburg in western Siberia and the construction of the associated "Progress" pipeline network to transmit this gas to markets in European Russia and Eastern and Western Europe. To varying degrees, all East European states are participants in this undertaking, for which they will receive in compensation 20 to 22 billion cubic meters of natural gas annually for a twenty-year period beginning in 1989.[60] These annual imports are approximately equivalent to total 1985 regional production of natural gas, excluding the output of Romania. Romania and Czechoslovakia, which are making the most extensive contributions, are due to receive in return the largest volumes of natural gas.[61]

East European participants have expressed considerable dissatisfaction with the joint projects.[62] In particular, they are not convinced that these projects represent the economically optimal method for solving their energy problems. One critic charged that the decision to pursue a particular project "largely constitutes a political commitment without any reference to cost/benefit analyses of the project's merit."[63] In effect, East European creditors also provide the Soviet debtor with a "hidden" subsidy through artificially low interest rates (usually 2 percent) on credits in joint projects. What is more, they complain, the CMEA practice of valuing credits and payments at their respective times of delivery has the result of devaluation in times of inflation. In turn, the USSR criticizes the East Europeans for not meeting many of their obligations under the joint projects and for running huge trade deficits with the Soviet Union. A Soviet source contends: "When most European CMEA states are permanently indebted to the Soviet Union as a result of clearing their commodity trade, their deliveries to the Soviet Union under special-purpose credits no longer make any economic sense."[64]

What these reservations about joint projects portend for future deliveries of Soviet natural gas to Eastern Europe remains to be seen. The contracts now in force, if fulfilled, should lead to a modest increase through the year 2000 in the role of natural gas in regional primary energy consumption, which unquestionably would benefit the environment—especially because several governments indicate that highly polluted areas will receive priority in the use of natural gas.[65] Yet several factors limit the overall contribution that natural gas can make to improve the quality of the environment. First, only in Hungary and Romania is natural gas likely to constitute more than

20 percent of primary energy consumption by the year 2000. Second, limited resources for investment in all East European states may impede the sometimes costly process of reequipping enterprises to use natural gas instead of coal or oil in their production processes. Finally, in certain instances—motor vehicles are an obvious example—it simply is not technologically feasible or cost-effective to substitute natural gas for more polluting fuels.

Nuclear Energy

Prevailing opinion in Eastern Europe traditionally has viewed nuclear power as less threatening to the environment than other forms of primary energy, especially coal. As a Polish source explains, the choice is a simple one: if Poland were to forgo nuclear power and continue to rely upon coal, then the future would witness "irreversible destruction of our environment and the turning of Poland into the dirtiest and most poisoned country in Europe."[66] Similar attitudes are frequently expressed in other East European countries, notably Czechoslovakia and the GDR, explaining why many East European environmentalists voice little opposition to nuclear power.

Of course, the recent accident at the Chernobyl nuclear power station in the USSR, which caused widespread environmental contamination in parts of Eastern Europe (especially Poland), has cast doubt on this rationale for nuclear power. The accident provoked public protests against nuclear power, most notably in the GDR and Poland and, to a lesser extent, Czechoslovakia. The protesters questioned the environmental safety of nuclear power and pointedly criticized their governments' reticence to discuss openly both the magnitude and the consequences of the accident and the overall dangers that nuclear power poses to the population.[67]

These protests do not appear, at least publicly, to have shaken the official commitment of East European governments to develop nuclear power, although all governments now acknowledge the need to draw appropriate lessons from the Chernobyl accident. The nuclear power plans announced at the 1986 meeting of the CMEA Session of the Council in Bucharest reflect the continuing commitment to nuclear power. There, Nikolai Ryzhkov, chairman of the USSR Council of Ministers, announced that by the year 2000 East European states would expand their installed nuclear capacity to 50 million kilowatts, up from 8 million kilowatts of installed capacity in 1986.[68] At present, four states (Bulgaria, Czechoslovakia, GDR, and Hungary) operate nuclear power stations, and all states in the region have more of them under construction.[69] As in the past, however, plans do not involve construction of the Chernobyl-type graphite-moderated channel reactors; all Soviet-designed reactors will most likely be of the pressurized-water type similar to those commonly used in the West. The new reactors will probably be outfitted with external concrete containment structures, which are generally used at Western facilities but were absent from earlier East European models. (Table 2.3 provides data on the current and projected status of nuclear power in the region.)

The commitment to nuclear power appears strongest in Bulgaria and Czechoslovakia. The percentage of electricity generated by nuclear power in Bulgaria is among the six highest in the world, and the highest in Eastern Europe. In installed capacity of 1,760 megawatts, however, it ranks behind Czechoslovakia and the GDR. Nuclear power generated more than 30 percent of total domestically produced electricity in Bulgaria in 1986, and officials hope to increase production to 60 percent by the year 2000. In addition, both the existing Kozloduy nuclear power station and another one under construction at Belene are to provide for the centralized supply of heat to nearby population and industrial centers, and construction will soon begin on the country's first nuclear power station devoted solely to the production of heat.

Czechoslovakia had the largest installed nuclear capacity in Eastern Europe at the end of 1986 (2,640 megawatts), with two nuclear power stations currently in commercial operation: one at Jaslovské Bohunice

Table 2.3
ELECTRICITY OUTPUT FROM NUCLEAR POWER IN EAST EUROPEAN COUNTRIES
(Percent), 1985–2000

Country	Percent		
	1985–86	*1990*[a]	*2000*[a]
Bulgaria	32	44	60
Czechoslovakia	20	30	50
GDR	11	15	30
Hungary	25	NA[b]	40
Poland	—	9	40
Romania	—	18	NA[b]

SOURCES: Official publications of the respective countries.

[a]These projections should be treated cautiously since all countries listed have a record of substantial underfulfillment of their nuclear plans.

[b]Not available.

and the other at Dukovany in southern Moravia. By the year 2000, Czechoslovakia plans to have an additional reactor at Dukovany, as well as four more reactors at each of the two nuclear power stations currently under construction: at Mochovce in western Slovakia and Temelin in southern Bohemia. If all these plans materialize, Czechoslovakia will have 9,000 megawatts of installed nuclear capacity by the turn of the century. Nuclear power is projected to generate approximately 30 percent of all electricity in 1990 and as much as 50 percent by 2000.

The status of nuclear power is less certain in the GDR and Hungary. The GDR traditionally had been the leader in the development of nuclear power in Eastern Europe. The small power plant at Rheinsberg, operative in 1966, was the first commercially operating nuclear facility in the region. In 1974–79, four more reactors went into operation at its second nuclear power station, the Lubmin facility on the Baltic Coast. Thus, with 1,840 megawatts by 1980, the GDR possessed the largest installed nuclear capacity in Eastern Europe, generating 12 percent of all electricity produced in the country.

The GDR's political leadership has, however, apparently reevaluated the extent of its commitment to nuclear power. Erich Honecker himself has expressed the personal conviction that "atomic energy is not the last word," saying he was pleased that the GDR still relied for its power primarily on coal.[70] In fact, no new nuclear capacities were commissioned in 1981–85, and the percentage of electricity generated by nuclear power actually declined over the period. The national economic plan for 1986–90 now projects that by 1990 nuclear power will account for no more than 15 percent of the electricity generated in the GDR. Nonetheless, as reserves of brown coal approach depletion in the next century, the GDR may well be forced to revive its long-delayed commitment to construction of additional nuclear power plants.[71]

Hungary currently operates two 440-megawatt reactors at Paks on the Danube River. The Paks complex has incurred enormous cost overruns and interminable delays in construction, and two additional reactors projected for the site are still not operational. Government agencies "negotiated a long time" over the future expansion of Paks, specifically over whether there should be any expansion at all and, if so, whether it should involve reactors of 440-megawatt capacity or, as originally planned, 1,000-megawatt (MW) capacity. Finally, in August 1986 officials announced that the Soviet Union would give Hungary "technical assistance in the design and installation of two power units with 1,000 MW reactors" to be commissioned at Paks in the mid-1990s.[72]

Poland and Romania are the only two states in Eastern Europe without a commercially operative nuclear power station. The saga of nuclear power in Poland has been one of continually frustrated plans, dating back to 1971 when the government first indicated its intention to build a nuclear power station. Little was heard of this project until 1982, when the government suddenly announced the intention to build its first nuclear power station at Żarnowiec. Plans call for four reactor blocks with a total capacity of 1,860 megawatts, the first to be made operational in 1991 and the entire project to be completed by the late 1990s. Longer range plans call for the construction of two more nuclear power stations, each equipped with four Soviet-supplied 1,000-megawatt reactors, to be operating by the turn of the century. Polish sources themselves remain highly skeptical, however, that all or even a substantial part of these plans will materialize.[73]

The nuclear power program in Romania displays features bordering on the bizarre. These include the supervision of the program by members of the family of Nicolae Ceauşescu, general secretary of the Romanian Communist Party, who possess no appropriate scientific or technical training for this task; the promulgation and repeated revision of unrealistic and unfulfilled targets; and groundless claims about Romania's ability to produce nuclear equipment, including nuclear weapons, for domestic use and for export. Another unusual feature of the Romanian program involves plans to import nuclear technology from the West, including Canadian-built reactors. Hard-currency shortages may make realization of these plans difficult.

On balance, nuclear power probably will remain an important, albeit not the decisive, component of the energy balance in the region. The record to date includes extended delays, enormous cost overruns, frequently revised and sometimes abandoned plans, and a plethora of seemingly insoluble logistical, organizational, and administrative problems. Ironically, however, even after Chernobyl, concern for environmental quality may remain one of the strongest rationales for nuclear power in Eastern Europe.

PROSPECTS

East European planners have pursued policies that foster the excessive consumption of highly polluting fuels, thereby contributing (however unintentionally) to massive degradation of the environment in many parts of the region. Reductions in the overall consumption of energy, and in particular of solid fuels, are essential to efforts to improve the quality of the environment.

One cannot be sanguine about the prospects for achieving meaningful reductions in the consumption of energy. To realize such reductions would require altering or eliminating those features of the "command economy" and the associated strategy of "extensive" growth that continue to promote excessive consumption of energy throughout the region. Most fundamentally, the persistent bias in these countries toward supply-side initiatives to expand production, rather than demand-side initiatives to limit consumption, must be remedied. Shifting away from this supply-side orientation will prove difficult. First, politically well connected bureaucrats with a vested interest in its perpetuation face little or no opposition from any comparably powerful constituency for energy conservation. Second, even if such a lobby existed, the enormous "sunken costs" arising from the operation of existing plants, together with the long lead time required to construct new power plants, effectively preclude substantial changes in the allocation of investment resources in the short run. Third, in an era of fiscal stringency it may prove difficult to justify the initial expenditures required for energy conservation projects the economic payoff of which may not be realized for many years.

Planners also must increase prices for energy resources to reflect their real marginal cost and thereby provide consumers with a positive economic incentive to conserve energy. To be effective, such a "market" approach to energy conservation must be accompanied by broader "marketization" of the national economy. Prices must reflect the actual relation of supply and demand. Although economic reform along these lines is now widely discussed in the context of Gorbachev's *perestroika* (economic restructuring), and has been in place to a limited extent for some time in Hungary, its adoption and implementation face powerful obstacles. Apart from posing a threat to the preeminence of the ruling Communist parties, the "market" approach also raises the politically sensitive question of increasing consumer prices to reflect the full value of energy contained in sundry goods and services. So far, elites have typically been reluctant to compromise any further their already precarious political legitimacy by enacting measures that, although they might promote the long-run welfare of the national economy, would redound to the short-run detriment of consumer welfare.

Finally, East European economies will remain extraordinarily energy-intensive so long as traditional core industrial sectors (steel, chemicals, mining, and the like) continue to be the primary consumers of energy. Even under optimal circumstances, a rapid change in production profile to emphasize less energy-intensive goods and services would involve enormous costs. Conditions for such a retooling in Eastern Europe are hardly favorable: requisite investment resources (especially of hard currency) are in exceedingly short supply, and politically painful dislocations (for example, increases in levels of unemployment) must inevitably accompany any major restructuring of the national economy.

The prospects for a substantial diminution in reliance on solid fuels also appear limited. These prospects were bright in the 1960s and early 1970s as large volumes of cheap Soviet oil began to replace coal in primary energy consumption, and as natural gas and nuclear power came to assume a greater, albeit still limited, role in the energy balance of the region.

The twin oil shocks of 1973 and (especially) 1979 halted this trend. The dramatic decline in world oil prices since 1982 has not yet offset the shift away from oil, however, for two reasons: first, Eastern Europe possesses insufficient reserves of hard currency to buy large amounts of even relatively cheap oil on the world market; and second, the USSR has limited the volume of oil for export to Eastern Europe while demanding more hard goods in payment.

The considerable difficulties encountered in expanding domestic production of coal in 1981–85 appear to have tempered the more grandiose dreams of planners (except in Romania) who initially saw increased reliance on coal as the unavoidable response to higher oil prices. Nevertheless, coal may be expected to retain or increase somewhat its share in all energy balances of the region, except in Hungary and perhaps Czechoslovakia, through the remainder of this century. That this coal typically is of diminishing caloric and increasing polluting content represents one of the gravest threats to the future of the environment in Eastern Europe.

In sum, present and likely future trends in energy policies of the region make optimism about environmental prospects difficult. Ironically, grounds for pessimism may increase if the stagnant economies of most East European states begin to revive, because the resulting boost to energy consumption would most probably worsen pollution in the region in the absence of additional measures.

NOTES

1. In this study "Eastern Europe" refers to Bulgaria, Czechoslovakia, the GDR, Hungary, Poland, and Romania. The term "Communist Europe" encompasses the states of Eastern Europe and the USSR.
2. *Kulgázdaság* (Budapest) 1 (January 1984), in Joint Publications Research Service-*Eastern Europe Industrial and Economic Affairs* (hereafter JPRS-EEI), 13 March 1984, p. 13; *Kulpolitika* (Budapest) 3 (1985), in JPRS-EEI, 30 August 1985, p. 15.

3. *Héti Vilaggázdaság* (Budapest), 11 October 1986, in JPRS-*East Europe Report* (hereafter JPRS-EER), 6 March 1987, p. 3.
4. For a detailed exposition of this argument, see *Figyelő* (Budapest), 13 October 1983, in JPRS-EEI, 6 December 1983, p. 8.
5. Romania possesses more than 80 percent and 40 percent, respectively, of the oil and natural gas reserves in the region. Poland possesses more than 90 percent of the reserves of hard coal.
6. For an overview of the environmental situation in Eastern Europe, see John M. Kramer, "Environmental Problems in Eastern Europe: The Price for Progress," *Slavic Review* (summer 1983).
7. For example, a Soviet source in 1987 reports that the European members of CMEA "consume approximately 40 percent more power per unit of output on average than European Economic Community countries" (*New Times* [Moscow], 12 January 1987, p. 32). Other analyses expressing approximately similar conclusions include *Pogled* (Sofia), 11 November 1985, in JPRS-EER, 16 April 1986, p. 5; *Rude Pravo* (Prague), 16 June 1986, in Foreign Broadcast Information Service-*Daily Report: Eastern Europe* (hereafter FBIS-EEU), 23 June 1986, p. D5; *Kommunist* (Moscow), May 1983, pp. 73–84.
8. Data from Leslie Dienes and Viktor Merkin, "Energy Policy and Conservation in Eastern Europe," in Joint Economic Committee of the U.S. Congress, *East European Economies: Slow Growth in the 1980's*, vol. 1 (Washington, D.C.: Government Printing Office, 1985), p. 351 (hereafter cited as "Energy Policy and Conservation").
9. For an elaboration of this point, see, for example, *Figyelő*, 27 June 1979, quoted in Radio Free Europe (RFE) *Situation Report* (Hungary) 14 (18 July 1979); *Rude Pravo*, 18 November 1981, in FBIS-EEU, 23 November 1981, p. D2; *Rabotnichesko delo* (Sofia), 7 October 1985, in JPRS-EER, 21 November 1985, p. 40.
10. *Pogled*, 8 July 1985, in JPRS-EEI, 1 October 1985, p. 27.
11. For a typical exposition of this argument, see *Przeglad techniczny* (Warsaw), 29 April 1984, in JPRS-EEI, 23 July 1984, p. 114.
12. *Pogled*, 11 November 1985; *Energieanwendung* (Leipzig), 5 (September-October 1983), in JPRS-EEI, 16 December 1983, p. 55.
13. *Pravda* (Bratislava), 23 October 1984, in JPRS-EEI, 16 November 1984, p. 1; *Rabotnichesko delo*, 1 July 1985, in JPRS-EEI, 3 September 1985, p. 19; *Voprosy ekonomiki* 12 (December 1980): 101.
14. On this subject see *Rude Pravo*, 21 September 1985, in FBIS-EEU, 1 October 1985, p. D5; *Hospodářské noviny*, 3 August 1984, in JPRS-EEI, 3 October 1984, p. 23.
15. See Istvan Dobozi, "The 'Invisible' Source of 'Alternative' Energy: Comparing Energy Conservation of the East and West," *Natural Resource Forum* 3 (1983): 213 (hereafter cited as " 'Invisible' Source"). Analysis of the national economic plans for 1981–85 from *Voprosy ekonomiki* 8 (August 1983): 105–13. On this subject also see *Voprosy ekonomiki* 12 (December 1980): 99.
16. See Dienes and Merkin, "Energy Policy and Conservation," especially pp. 337–55, for a discussion of these programs.
17. *Hospodářské noviny* (Prague), 10 August 1984, in JPRS-EEI, 21 September 1984, p. 6; *Svet hospodářství* (Prague), 22 September 1983, in JPRS-EEI, 23 November 1983, p. 17; Dobozi, " 'Invisible' Source," p. 213.
18. For example, Hungarian sources estimate that lower than anticipated economic growth, especially in energy-intensive industries, accounts for as much as two-thirds of the savings in energy realized in 1981–85. *Népszabadság* (Budapest), 16 October 1985, in FBIS-EEU, 23 October 1985, pp. F7–8.
19. *Magyarország* (Budapest), 7 July 1985, in JPRS-EEI, 7 September 1985, pp. 28–29.
20. Data in table 9 from Central Intelligence Agency (CIA), *Energy Supplies in Eastern Europe: A Statistical Compilation*, ER79-10624, December 1979 (hereafter cited as *Energy Supplies in Eastern Europe*).
21. Thus a Soviet source estimates that between 1971 and 1978 on average the cost

of extracting one ton of coal doubled in the region. This source also reports that in Czechoslovakia the cost of extraction per ton of coal was 60 percent higher in 1976–80 than in 1971–75. *Voprosy ekonomiki* 12 (1980): 98.

22. For a discussion of the impact on Eastern Europe of escalating world market and Soviet prices for oil after 1973, see John M. Kramer, "Between Scylla and Charybdis: The Politics of Eastern Europe's Energy Problem," *Orbis* (winter 1979): 936–44.

23. Data derived from tables 104 and 105 in Directorate of Intelligence, *Handbook of Economic Statistics, 1986* (Washington, D.C.: Government Printing Office, 1986).

24. See *Trybuna ludu* (Warsaw), 3 January 1981, cited in RFE *Situation Report* (Poland), 16 January 1981, for an admission of this circumstance. For an analysis of why most countries failed to fulfill their plan targets for coal, see *Figyelő*, 13 October 1983, in JPRS-EEI, 6 December 1983, p. 8.

25. For details of these plans, see *Agerpres* (Bucharest), 14 July 1981, in FBIS-EEU, 15 July 1981, p. H7; *Bucharest Domestic Service*, 18 March 1982, in FBIS-EEU, 22 March 1982, p. H13.

26. Data on plans for production in 1985 from *Hétőfi Hirek* (Budapest), 3 August 1985, in JPRS-EEI, 27 September 1985, p. 22. Data on financial investments in the coal industry from *Figyelo*, 19 September 1985, in JPRS-EEI, 21 November 1985.

27. Directorate of Intelligence, *Handbook of Economic Statistics, 1986*, tables 104 and 105.

28. For data on these plans for Bulgaria, the GDR, Poland, and Romania, see, respectively, *Energetika* (Sofia), 2 (1985), in JPRS-EEI, 29 July 1985; *Neues Deutschland* (East Berlin), 23 April 1986, in FBIS-EEU, 10 June 1986; *Zycie Warszawy*, 9 July 1985, in JPRS-EEI, 26 August 1985, p. 62; *Agerpres*, 5 June 1984, in FBIS-EER, 6 June 1984, p. H6.

29. PAP (Warsaw), 7 August 1986, in FBIS-EER, 8 August 1986, p. G20.

30. *Hospodářské noviny* 34 (1986), in JPRS-EER, 10 December 1986, p. 35.

31. For a detailed discussion of this issue, see *Hospodářské noviny* 43 (1985), in JPRS-EER, 21 January 1986, p. 3.

32. Data from Stanley Kabala, "Poland: Facing the Hidden Costs of Development," *Environment* 9 (1985): 11.

33. Estimates on increases in sulfur dioxide emissions from Polish scientists as reported in *Dagens Nyheter* (Stockholm), 4 December 1981, in JPRS-EEI, 28 January 1982, p. 19. For the assessment that expanded mining operations may result in "total environmental degradation in the affected areas," see *Przegląd techniczny*, 29 April 1984, in JPRS-EEI, 23 July 1984, p. 118.

34. Data for Budapest from *Magyar Hirlap* (Budapest), 24 July 1984, in JPRS-EEI, 30 August 1984, p. 107; data for Warsaw from *Przegląd morski* (Warsaw) 4 (1981) in JPRS-EEI, 5 May 1982, p. 50.

35. See Jan Vanous, "Long-Term Trends in Energy Consumption and Trade in the Soviet Union and Eastern Europe," *Plan Econ Report* (31 March 1986), especially pp. 28–30.

36. On the origins and construction of the "Druzhba" pipeline, see RFE *Situation Report* (Czechoslovakia), 9 February 1977.

37. Data from table 9 in CIA, *Energy Supplies in Eastern Europe*.

38. *Neues Deutschland*, 21 April 1981; *Ekonomicheskaia gazeta* (Moscow) 45 (1981).

39. See John M. Kramer, "Soviet-CMEA Energy Ties," *Problems of Communism* (July–August 1985), especially pp. 36–37, for a detailed analysis of the Soviet action.

40. See *Pravda* (Moscow), 3 April 1984, for a frank exposition of problems in the Soviet petroleum industry. A detailed analysis of this subject by a Hungarian scholar appears in *Kulgázdaság* 1 (January 1984), in JPRS-EEI, 13 March 1984, p. 13.

41. *Wall Street Journal*, 16 March 1983; *Oil and Gas Journal*, 13 March 1985. These data should be taken as approximations, because the USSR since 1977 has ceased publishing information on the quantity of its energy exports, giving only value data.

42. See *Kulgázdaság* 1 (January 1984), in JPRS-EEI, 13 March 1984, p. 17. See also

Edward Hewitt, "Soviet Primary Products Export to Comecon and the West," in *Soviet Natural Resources in the World Economy*, R. G. Jensen, ed. (Chicago: University of Chicago Press, 1983), pp. 639–57.

43. On the "Bucharest" formula, see RFE *Background Report* (Eastern Europe), 25 March 1977. *New York Times*, 25 January 1975, provides details of the revised price formula.
44. RFE *Background Report* (Economics), 24 August 1984. These data are calculated from dollar/ruble exchange rates devised by Hungarian foreign trade experts that provide a more realistic price for Soviet oil than does the "official" CMEA price, which is based on the artificially high official rate of exchange. In 1984, the "official" CMEA price for oil actually approximated—and in subsequent years has exceeded—the comparable world-market price.
45. A frank admission of this circumstance is carried by the official Polish news agency PAP, 23 September 1981, in FBIS-EEU, 24 September 1981, p. G3.
46. *Pravda* (Moscow), 9 October 1983.
47. A Hungarian commentary explicitly identified this nexus between political and economic stability: "The events in Poland show that the malfunction of the economic system can lead to a serious social crisis. Economic problems obviously affect the atmosphere of society and even the stability of the political system." *Budapest Domestic Service*, 28 April 1981, in FBIS-EEU, 1 May 1981, p. F3.
48. *Rude Pravo*, 19 October 1983.
49. *Magyarorszag*, 7 July 1985, in JRPS-EEI, 7 September 1985, p. 30.
50. See PAP, 20 March 1985, in FBIS-EEU, 21 March 1985, p. G4; *Prague Domestic Service*, 6 February 1986, in FBIS-EEU, 10 February 1986, p. D6; *Magyar Hirlap*, 6 April 1986, in FBIS-EEU, 15 April 1986, p. F2.
51. Data on Hungary from *Népszabadság*, 16 October 1985, in FBIS-EEU, 23 October 1985, pp. F7–8.
52. *Przegląd techniczny*, 29 April 1984, in JPRS-EEI, 23 July 1984, p. 117.
53. For materials on this see, for example, *Népszabadság*, 20 October 1978, in JPRS-EEI, 1 December 1978, p. 8; *Rude Pravo*, 3 September 1980; *DIW Wochenbericht* (West Berlin), 52 (25 July 1985), in JPRS-EER, 17 January 1986, p. 41.
54. Data on production of natural gas from Directorate of Intelligence, *Handbook of Economic Statistics, 1986*, table 103. Data on consumption from table 9 in CIA, *Energy Supplies in Eastern Europe*.
55. See Edward Hewett, "The Near-Term Prospects for Soviet Natural Gas Industry," in Joint Economic Committee of the U.S. Congress, *Soviet Economy in the 1980's: Problems and Prospects* (Washington, D.C.: Government Printing Office, 1982), pp. 391–413.
56. *Pravda* (Moscow), 27 June 1979; U.S. Congress, Office of Technology Assessment, *Technology and Soviet Energy Availability* (Washington, D.C.: Government Printing Office, 1981), pp. 286–90.
57. See John Hannigan and Carl McMillan, "Joint Investment in Resource Development: Sectoral Approaches to Socialist Integration," in Joint Economic Committee of the U.S. Congress, *East European Economic Assessment*, vol. 2 (Washington, D.C.: Government Printing Office, 1981), pp. 259–95. For an analysis of joint projects specifically related to the energy sector, see John M. Kramer, "Soviet-CMEA Energy Ties," *Problems of Communism* (July–August 1985), pp. 45–46.
58. See RFE *Background Report* (Eastern Europe), 2 December 1975.
59. For the estimate, see RFE *Situation Report* (Czechoslovakia), 19 May 1983. On the participation of the East European states, see *Rude Pravo*, 18 June 1982 and 4 August 1983; *Hétöfi Hirek*, 22 February 1982, in FBIS-EEU, 15 March 1982, p. F5; *Politika* (Belgrade), 31 October 1982, in FBIS-EEU, 5 November 1982, p. 19.
60. See *Pravda* (Moscow), 30 July 1984, for details of the project.
61. Materials on Romania and Czechoslovakia from, respectively, *Agerpres*, 24 March 1986, as summarized in RFE *Background Report* (Eastern Europe), 15 July 1986; *Prague Television Service*, 9 January 1986, in FBIS-EEU, 22 January 1986, p. D8.

For materials on the participation of other East European states in this undertaking, see *Rabotnichesko delo* (Moscow), 22 January 1986, in JPRS-EER, 25 April 1986, p. 20; *Izvestiia* (Moscow), 22 January 1986, in FBIS-*Daily Report: Soviet Union*, 23 January 1986, p. F2; *Magyar Hirlap*, 31 January 1986, in FBIS-EEU, 27 February 1986, p. F1.

62. A detailed exposition of their complaints is in *Kulgázdaság* 1 (January 1984), in JPRS-EEI, 13 March 1984, pp. 45–46.
63. Kalman Pecsi, *The Future of Socialist Integration* (New York: M. E. Sharpe, 1981), quoted in RFE *Background Report* (Eastern Europe), 10 December 1982, p. 5.
64. *Planovoe khoziaistvo* (Moscow), August 1981, p. 19.
65. For example, a Czechoslovak source reports that in the use of natural gas there is an "unconditional necessity to give preference to selected territories with the most rapidly deteriorating environments." *Planované hospodářství* 4 (Prague: 1984), in JPRS-EEI, 5 September 1984, p. 50. For the expression of similar sentiments in the GDR, see *East Berlin Domestic Service*, 15 June 1986, in JPRS-EER, 8 August 1986, p. 23.
66. *Rzeczpospolita* (Warsaw), 28 December 1984, in JPRS-*Political, Sociological, and Military Affairs*, 26 February 1985, p. 191.
67. For further details on these protests, see John M. Kramer, "Chernobyl and Eastern Europe," *Problems of Communism* (November–December 1986), pp. 41–43.
68. *Pravda* (Moscow), 4 November 1986.
69. For a detailed discussion of the development of nuclear power in Eastern Europe, see *Mezhdunarodnoe sotrudnichestvo stran chlenov SEV v oblasti atomnoi energetiki* (Moscow: Energoatomizdat, 1986): 56–105. Unless otherwise noted, all materials on nuclear power in the present paper are drawn from this source.
70. Honecker's remarks, made in an interview to *Dagens Nyheter*, were reprinted in *Neues Deutschland*, 25 June 1986.
71. See other parts of this volume for a more detailed description of GDR's energy policy.
72. See *Népszabadság*, 18 October 1985, in JPRS-EEI, 12 December 1985, p. 76, for information regarding debates on the future expansion of Paks. *Izvestiia*, 16 August 1986, reports the agreement for 1,000-megawatt reactors.
73. See, for example, the assessment in *Zycie Warszawy*, 10 October 1983, in JPRS-EEI, 5 December 1983, p. 31.

II

THE INTERNATIONAL CONTEXT

3

ENVIRONMENTAL PROTECTION IN SOVIET-EAST EUROPEAN RELATIONS

Charles E. Ziegler

The Soviet Union in the late 1980s is more closely integrated into the world community than at any time in the past. This new interdependence extends beyond politics and economics into the field of environmental protection. At the Twenty-seventh Party Congress in February–March 1986, Mikhail Gorbachev referred to enviromental pollution and the depletion of natural resources as global problems "affecting the very foundations of the existence of civilization." Environmental degradation, he asserted, "cannot be resolved by one state or a group of states. This calls for cooperation on a worldwide scale, for close and constructive joint action by the majority of countries." Effective international procedures and mechanisms are needed to counteract the "ill effects of exposing nature to the blind play of market forces."[1] In line with these statements, the general pattern of East-West collaboration on environmental matters has improved since 1985.[2]

Important changes are also occurring domestically in Soviet environmental protection under Gorbachev. Although the focus of environmental efforts has not shifted perceptibly (water protection remains the single most important issue, absorbing close to three-fourths of total capital investment in environmental protection measures), the process by which these issues are addressed has changed markedly. First, in keeping with the new policy of *glasnost* (openness), environmental questions are being debated more openly and honestly than at any time since the initial furor over Lake Baikal in the mid-1960s.

Second, the participation of nonspecialists, particularly literary figures, in these debates has become routine. Moreover, Soviet writers almost invariably adopt environmentalist positions, and they are not reluctant to couch their criticisms of developers in highly emotional terms. One of the most significant aspects of such emotive appeals is the nationalistic tone that has characterized recent debates. This nationalism is frequently Russian (as in the river diversion debate), but links between ecological concern and ethnic sentiment are evident in other republics as well. Finally, the highest levels of party and government seem today to be more willing to respond to environmental crises than at any time in the past. Despite considerable disagreement over many of Gorbachev's reform initiatives, the Soviet leadership appears to be in basic accord on the need to reduce costs resulting from environmental degradation. Tough decisions that had been postponed in recent years are now being addressed decisively.[3]

The domestic ecological debate in the USSR reflects a changing political environment, which also involves new approaches to fundamental economic problems. Ecological problems are increasingly viewed as symptoms of economic inefficiency, as one factor contributing to the general economic malaise. Many of the wasteful policies of the Brezhnev era regarding the use of resources and energy and environmental protection are no longer considered economically feasible. Soviet economic relations with Eastern Europe are also being affected by Gorbachev's domestic initiatives. Policies that were, from the Soviet perspective, either wasteful or overgenerous (such as supplying oil at favorable prices and buying second-rate finished goods from the East Europeans) are being further modified to pressure the East Europeans to bear greater responsibility for making the socialist community more efficient. Ecological problems, as a subset of the larger economic questions of modernization and improved efficiency within the Council for Mutual Economic Assistance (CMEA),[4] are likely to be affected by these new directions in Soviet policy.

Eastern Europe has the dubious distinction of being one of the most heavily polluted and resource-wasteful regions in the world. In an era of economic crisis for the USSR and its allies, the substantial costs generated by pollution and waste of natural resources can no longer be tolerated. Closer collaboration among CMEA member states on environmental questions coincides with the broader Soviet strategy of improving scientific and technical cooperation within the alliance, modernizing the Soviet and East European economies, and accelerating the shift to intensive development.

This chapter outlines the structures that play a role in environmental cooperation among the CMEA member states and examines the circumstances that condition members' efforts. While pollution problems and the exploitation of natural resources are aggravated by the Stalinist centralized economic system, the political constraints of the alliance also inhibit effective multilateral approaches to solving these problems.

THE USSR AND EASTERN EUROPE

The first stage of CMEA cooperation on environmental protection dates from the early 1960s. A 1962 resolution of the council mandated the coordination of scientific and technical research on air and water pollution, but efforts in this direction were minimal before 1974.[5] In that year the member states signed an agreement on further scientific and technical cooperation for developing environmental protection measures and created a Joint Council for the Protection and Improvement of the Environment to coordinate their efforts. According to CMEA statutes, this council is responsible for coordinating scientific and technical efforts in the area of environmental protection and rational use of resources, advancing proposals for pollution standards among member states, assisting member states in information exchanges, promoting specialization in the production of pollution abatement equipment, and forecasting and analyzing the state of the environment. Yugoslavia, an associate member of CMEA, is also a member of the joint council.[6]

In 1974 the CMEA executive committee enacted a comprehensive program on environmental protection outlining measures in twelve areas of concern: socioeconomic, organizational-legal, and pedagogical aspects of environmental protection; hygiene; protection of the ecosystem and landscape; air pollution; meteorological aspects of atmospheric pollution; noise pollution; water pollution; treatment of municipal, industrial, and agricultural wastes; radiation; protection of mineral resources and the rational use of natural resources in general; city planning and population settlements; and the development of low-waste and waste-free technologies. In 1976–80, more than 540 organizations cooperated on some two thousand technical processes and methodological questions. The CMEA cooperation program for 1981–85 was expanded to add global environmental monitoring and informational problems to the original twelve areas of cooperation.[7]

According to one Soviet author, the comprehensive program for economic integration among CMEA members, adopted in 1971, "gave new impetus to the cooperation between the CMEA countries in protecting the natural environment."[8] It is generally acknowledged, however, that integration of the community has proceeded slowly in most areas. One Western specialist perceives the CMEA member states as sincere in their intentions about integration but lacking precisely defined and generally accepted principles for implementing their goals.[9]

WATER SUPPLY AND POLLUTION

Much of the CMEA cooperation on environmental matters has been in the field of water supply and water pollution, which was singled out as an urgent problem at the 1971 conference that led to the adoption later that year of the comprehensive program. The East European countries suffer from water problems that are, if anything, more acute than those in the European USSR. Since 1962 the heads of water supply organizations in the various CMEA countries and Yugoslavia have collaborated on scientific and technical questions relating to water usage and pollution.

The East European countries vary considerably in physical size, population density, level of industrialization and urbanization, water supply, and seriousness of water pollution problems. According to one Soviet study, the German Democratic Republic (GDR) has the most severely polluted water, followed by Hungary and Bulgaria in the second category, Czechoslovakia and Poland in the third, and finally Romania and Yugoslavia. The GDR, Czechoslovakia, and Poland—countries that experienced relatively early industrialization—have the largest number of small, older plants. In 1979, for example, 65 percent of the GDR's enterprises had 200 or fewer employees, while only 11 percent had 1,000 or more workers. The small size of industrial enterprises in these countries makes it less cost-effective to install waste purification facilities. Yet the cost of using polluted water in the production process (and the GDR uses 69 percent of its total water budget for industry) can be as much as five times more expensive than using clean water.[10] To make matters worse, all of the East European countries, excluding Yugoslavia, are water-poor in comparison to Europe as a whole (see table 3.1).

The Soviet Union, Romania, Czechoslovakia, Hungary, and Yugoslavia have organized a permanent working group in the framework of a statute on the Meeting of Heads of Water Agencies to review problems of development in the Tisza River basin. Within the Tisza watershed, an area occupying 157,000 square kilometers, primary concerns have been flooding (a major flood damaged large areas of

Table 3.1

AVERAGE ANNUAL WATER RESOURCES AND WATER USE IN EUROPEAN SOCIALIST COUNTRIES

Country	River flow (cubic kilometers)	Water resources per square kilometer (thousands of cubic meters)	Water resources per capita (cubic meters)	Demand of all water users for water (cubic kilometers, per year, for 1975–80)	Level of use of water resources (percent)
Bulgaria	18.8	166	2,140	8.0	42
Czechoslovakia	27.4	214	1,800	5.6	21
Hungary	11.0	116	1,030	5.0	45
GDR	17.7	164	1,050	9.3	52
Poland	56.5	180	1,620	13.8	24
Romania	46.4	195	2,100	16.0	34
Yugoslavia	126.0	470	5,800	8.3	6
European USSR	1,009.0	203	6,000	NA[a]	NA[a]
Europe as a whole	3,112.0	319	4,800	NA[a]	NA[a]

SOURCE: E. M. Goncharova, "Vliianie vodno-ekologicheskikh uslovii narazvitie i razveshchenie promyshlennosti europeiskikh stran-chlenov SEV," in *Voprosy geografii: geografiia khoziaistva stran-chlenov SEV v usloviakh integratsii*, 123 (Moscow: Mysl', 1984), p. 301.
[a]Not available.

Hungary and Romania in spring 1970) and protecting the Tisza and its tributaries from pollution. The quality of water in the Tisza and its tributaries has declined in recent years as communal sewage (much of which is untreated) and agricultural runoff (nitrogen and phosphorus) have polluted the region's waterways. Hungary has taken a leading role in coordinating efforts to collect and disseminate hydrometeorological data.[11] The five nations have worked out a broad plan for transportation, fishing, and recreational use of the Tisza River and its tributaries.[12] Real progress toward effective coordination has been slow, but as one Hungarian official remarked, "undoubtedly the document worked out among the five countries can become a good example of multilateral cooperation on water economy by the socialist states."[13]

In 1973, nine CMEA member states (Bulgaria, Hungary, the GDR, Cuba, Mongolia, Poland, Romania, the USSR, and Czechoslovakia) established Vodoinform, an information clearinghouse on water questions. Vodoinform, one of twenty-three projects functioning under MOSNTI (the International Branch System of Scientific and Technical Information), has enabled the CMEA countries to set up a water information journal (based in Bratislava, Czechoslovakia), arrange exchanges of scientific films and water-quality surveys, and create an automated document search service. In 1985 this service conducted searches on 213 inquiries, drawing on a data base of some 14,300 written documents.[14]

Since the mid-1970s the CMEA countries have put into operation a number of automatic water-quality monitoring stations that analyze water temperature, sulfur content, pH, electrical conductivity, and water level. Cooperative efforts are used both in constructing new automated facilities and in modernizing existing systems. In recent years Soviet authorities have urged a more extensive division of labor among CMEA members in producing pollution control equipment.[15] This urging is consistent with the general Soviet pattern of encouraging greater specialization within the bloc to realize the benefits of comparative advantage. As one Western observer has noted, however, specialization within CMEA is likely to be administrative, that is, based on political considerations rather than genuine opportunity costs.[16]

One attempt to increase specialization in water protection has been the formation in 1977 of the *Intervodoochistka* (Joint Water Purification) Economic Association. The participating members—Bulgaria, Hungary, the GDR, Poland, Romania, Czechoslovakia, and the USSR— agreed, initially for a ten-year period, to broaden economic and scientific-technical cooperation in the areas of scientific research, project design, and production concerning the creation, introduction, and operation of equipment and installations for purifying wastewater. *Intervodoochistka*'s governing council, which sits in Sofia, is made up of representatives (one from each country) who nominate working groups of seventeen specialists each to address specific questions of water protection.[17] The observation of one Soviet author that "the first steps have already been taken" in the cooperative production of purification equipment seems to suggest, however, that integration in this field has been modest at best.[18]

It is important to recall that all the East European members of CMEA border either directly on the Soviet Union (Poland, Czechoslovakia, Hungary, Romania) or on major bodies of water that are part of Soviet frontiers (Bulgaria and the GDR). The Soviets have concluded both bilateral and multilateral agreements to regulate the use and protection of various waterways they share with individual countries or groups of countries. In 1971 the Soviet Union and Romania concluded a joint hydroelectric power project on the Prut River. Stations were constructed on both the Soviet and the Romanian sides of the border, and the countries assumed responsibility for maintaining water quality on their respective sides.[19] Other cooperative mechanisms include the 1964 Agreement of Poland and the USSR on the Water Economy of Frontier Waters; the 1964 Soviet-Finnish Agreement on Frontier Water Systems; the 1974 Helsinki Convention on Protection of the Baltic Sea (Poland, the USSR, the GDR, Sweden, Finland, Denmark, and the Federal Republic of Germany [FRG]); and the 1975 Agreement of CMEA with the Danube Commission on protecting the Danube water basin from pollution. Finally, some Soviet and East European cooperation on water pollution takes place through the United Nations Economic Commission for Europe and the United Nations Environment Program.[20]

AIR QUALITY

Eastern Europe suffers from air pollution of a far greater magnitude than does the USSR. Within the socialist community, the GDR, Czechoslovakia, Poland, and Yugoslavia are the most heavily polluted countries.[21] CMEA's Standing Commission on the Coordination of Scientific and Technical Research has organized research on methods of scrubbing sulfur anhydride from exhaust gases, using the gas and aerosol by-products of the chemical industry; improving methods for gas and dust scrubbing; studying the effects of air pollution on health; and determining maximum permissible concentrations of pollutants in the ambient air. CMEA efforts to combat air pollution are coordinated by the Institute for Atmospherics and Refrigeration Technics, in Dresden.[22] Member states met in 1982 to discuss the effects of transnational air pollution,[23] a problem that is becoming more acute as the economic and health costs of acid rain gain greater attention.

Five member nations of the CMEA—Bulgaria, Czechoslovakia, East Germany, Hungary, and the Soviet Union—are signatories to the "30 percent" protocol on reductions in national sulfur dioxide emissions. Poland apparently has not considered signing the protocol because the Polish government cannot afford to undertake serious measures to reduce sulfur dioxide emissions.[24] The CMEA countries are particularly concerned about these and nitrogen oxide emissions in Europe because the prevailing west-to-east wind patterns deposit substantial amounts of these pollutants from France, West Germany, Italy, Spain, and Great Britain on their territories. But sulfur dioxide emissions in Czechoslovakia, the GDR, and Poland are equal to or exceed emis-

Table 3.2
ANNUAL SULFUR DIOXIDE EMISSIONS AND NITROGEN OXIDE EMISSIONS
IN EUROPE AND THE UNITED STATES

Country	Sulfur dioxide emissions[a] Total (thousands of metric tons)	Per capita (kilograms)	Nitrogen oxide emissions[b] Total (thousands of metric tons)	Per capita (kilograms)
Austria	294	39	216	29
Belgium	612	62	385	39
Bulgaria	1,000	112	200	22
Canada	4,520	186	1,750	72
Czechoslovakia	3,250	211	1,120	73
Denmark	408	78	290	55
Finland	360	74	248	52
France	2,250	42	1,693	31
FRG	2,750	45	3,100	50
GDR	4,000	239	800	48
Greece	400	41	150	15
Hungary	1,650	154	300	28
Ireland	140	40	75	21
Italy	3,150	55	1,462	25
Luxembourg	28	76	22	60
Netherlands	340	24	480	34
Norway	100	24	138	35
Poland	4,100	112	840	23
Portugal	342	34	330	33
Spain	2,633	70	800	21
Sweden	330	40	305	36
Switzerland	86	13	214	33
UK	3,690	66	1,689	30
USA	20,800	90	19,700	85
USSR	11,800	74	2,930	18
Yugoslavia	1,176	52	190	8

SOURCE: "Cooperative Program for Monitoring and Evaluation of the Long-Range Transmission of Air Pollutants in Europe (EMEP)," in *Background Brief* (London: Foreign and Commonwealth Office, April 1987).
[a]Sulfur dioxide figures relate to 1983, the most recent year for which comparable figures are available for most countries.
[b]Nitrogen oxide figures are mostly for 1984 or 1985, but national figures are less reliable than for sulfur dioxide.

sions from the worst polluters in Western Europe (see table 3.2). Much of this pollution is deposited on Soviet territory—two million metric tons of sulfur dioxide alone, according to one estimate of the net

Table 3.3
SULFUR DIOXIDE EMISSIONS IN SELECTED COUNTRIES,
1982

Country	Emissions (kilograms per $1,000 GNP)
Market-oriented	
Japan[a]	1
Sweden	4
France	5
FRG	5
USA	7
UK	8
Canada[a]	18
Centrally planned	
USSR	19
Romania	28
Hungary	31
GDR	35
Czechoslovakia	40

SOURCE: Lester R. Brown et al., *State of the World 1987: A Worldwatch Institute Report on Progress Toward a Sustainable Society* (New York: W. W. Norton, 1987), p. 187.
[a]Data for 1980.

annual inflow.[25] (On a visit to the Soviet Union in 1987, I was assured that the figure was closer to *twelve* million tons.)

Although absolute levels of emissions and depositions are crucial in evaluating the damage inflicted by transboundary pollution, these figures tend to distort the comparative "environmental efficiency" of different systems. The Worldwatch Institute's calculation of sulfur dioxide emissions for selected market and centrally planned economies per dollar of gross national product (GNP) throws into stark relief the difference between the two systems in damage per unit of output (see table 3.3). The Worldwatch study suggests that energy-intensive economic strategies, a heavy dependence on coal in Eastern Europe, and a virtually complete lack of emissions controls account for the poor performance of the centrally planned economies.[26]

ENERGY RESOURCES AND RAW MATERIALS

The joint pollution problems of the Soviet Union and Eastern Europe are largely a result of their economic strategies and energy balance. The CMEA nations are extremely wasteful in their use of energy. In

1982, per capita consumption of energy in Eastern Europe was 28 barrels of oil equivalent (boe). By contrast, the West European countries in the European Economic Community averaged per capita energy consumption of 25 boe, with a per capita income 30 percent above that of Eastern Europe.[27] In the period from 1970 to 1983, the United States reduced its energy consumption by 30 percent per unit of output. The West European countries reduced their energy consumption by an average of 14 percent over the same period, while the USSR, Bulgaria, Romania, and Yugoslavia all *increased* their energy consumption per unit of output.[28]

Several factors have discouraged energy conservation in Eastern Europe. First, the Soviet model of industrialization imposed on Eastern Europe after World War II created economic structures that are inherently energy-intensive, wasteful, and inefficient. The centrally planned economies of Eastern Europe were designed to be autarkic rather than complementary, thus inhibiting movement toward a genuine division of labor. Second, the relatively low prices for oil supplied to the East Europeans by the Soviet Union in the 1970s reinforced a tendency to specialize in energy-intensive exports such as iron and steel products, petrochemicals, and refined oil products.[29] Third, with the drop in world oil prices in the early 1980s, CMEA's five-year rolling-average pricing mechanism effectively eliminated the Soviet "energy subsidy." Consequently some East European countries have been encouraged to rely increasingly on highly polluting domestic coal and at the same time to intensify efforts to conserve energy. Despite some limited successes with conservation, these efforts have not generally outweighed the detrimental effect on the environment brought by the shift from oil to coal. The outmoded industrial plant of both Eastern Europe and the USSR pollutes more heavily than newer equipment in Western countries, where the shift away from traditional industrial production in favor of newer service sectors has further reduced emissions.

On the other hand, some Soviet authors suggest that the USSR has shouldered much of the "ecological burden" for its East European allies. For example, Soviet scholars note that the Soviet share of hard coal, oil, and iron ore extracted within the alliance has increased substantially from 1950 to 1980 (the increases are from 64 to 68 percent, 88 to 98 percent, and 92 to 99 percent, respectively). The Soviet share of cellulose production has increased from 57 to 68 percent over the same period, and in electrical energy production their share has risen from 68 to 75 percent. This particular study asserts that by reducing the role of their extractive industries, the East European countries have preserved larger areas of plow land and standing timber.[30]

The Soviet view is that Eastern Europe has benefited substantially from Soviet supplies of cheap raw materials not subject to world-market price fluctuations. Environmentally, they argue, Soviet gas and oil saved Eastern Europe from having to rely on less efficacious local resources for energy, primarily brown coal and lignite. For example, one Soviet study calculates that each ton of imported oil saves the GDR from burning at least six tons of brown coal. In this view, because the GDR imports twenty-two million tons of oil (and six billion cubic meters of natural gas) annually, the environmental "savings" from burning cleaner fuels are tremendous. In addition, importing oil and gas reduces the environmental disruption from strip-mining coal—for each hundred million tons of brown coal mined in the GDR, approximately five hundred million cubic meters of earth must be displaced.[31]

These claims are profoundly insulting to many East Europeans, who argue that it was the Soviet Union that originally imposed the inefficient structure that is at the root of Eastern Europe's energy and ecological problems. Second, they point out, in exchange for "concessionary" deliveries of raw materials, the Soviets have frequently demanded "hard" goods from the East Europeans or have forced them to supply items below cost that are expensive to produce. Finally, Soviet–East European trade terms are becoming increasingly disadvantageous to the smaller East European states, as the USSR reduces energy supplies and demands higher quality goods in payment. At the same time, the weak state of the East European economies is forcing a reorientation away from economic cooperation with the West or with the Organization of Petroleum Exporting Countries, and toward closer cooperation with the USSR. The East Europeans have little choice but to expand economic ties to the Soviet Union in the near future.

Poland's situation illustrates the result of recent Soviet pressures for closer cooperation among the CMEA members and for an increased supply of better quality goods to the USSR from Eastern Europe. John Hardt and Jean Boone found that in the first six months of 1986 the value of Poland's fuel exports to the Soviet Union increased by 35.1 percent over the previous year. Fuel exports to countries outside the alliance, valuable as a source of much-needed hard currency, rose by only 0.5 percent over the same period. During the first nine months of 1986, the quantity of Polish coal exported to hard-currency countries declined by 15.7 percent, coke exports fell by 25.8 percent, and exports of refined petroleum products dropped by 20.5 percent. A similar pattern of trade was evident in metals and metal products—exports to socialist countries increased by 21.3 percent in the first six months of 1986, while exports to nonsocialist countries dropped 13.9 percent. The authors found something of a contradiction between Soviet and Polish priorities:

> Poland's modernization may require investment for energy conservation, while short term demands of the Soviet Union may require opening of new coal mines and steel mills to meet the post-Chernobyl needs of the Ukraine. The continued emphasis on investment in new coal mines and steel mills in current plans may thus be reinforced by Soviet self-interest in those projects.[32]

Such pressures not only place increased strains on the hard-currency trade balance of heavily indebted East European economies, but also restrict the ability of these countries to restructure economic development in directions that make them more economically competitive in international markets. Ecological considerations are clearly secondary under these conditions.

Nuclear Power

CMEA's energy constraints are a major factor contributing to the Soviet decision to continue its ambitious nuclear energy program in the wake of the Chernobyl disaster. The share of electrical output generated by nuclear power stations in the USSR was 10.8 percent in 1985; this figure is expected to increase to 20 percent by the year 2000. Comparable figures for the East European countries in 1985 were 31.5 percent for Bulgaria, 18.5 percent for Hungary, 14.6 percent for Czechoslovakia, and 9.8 percent for the GDR.[33] Poland and Romania had no plants in operation at that time. John M. Kramer points out that highly optimistic (and probably unrealistic) projections estimate that the share of Eastern Europe's electrical output produced by nuclear energy will double or even triple by the year 2000.[34] The chairman of the USSR State Committee on Atomic Energy Use, A. Petros'iants, identified twenty-four atomic power stations of the water-cooled type in operation in the USSR and Eastern Europe as of 1985, with about twenty additional units of the same type scheduled to be operational by 1990.[35] Soviet Premier Ryzhkov asserted at the November 1986 CMEA meeting that nuclear power production in socialist countries outside the USSR would increase from eight million to fifty million kilowatts by the year 2000.[36]

Despite all the problems associated with nuclear power, Eastern Europe's energy-poor situation makes nuclear power very difficult to reject as an alternative source of energy. Even after Chernobyl, nuclear power is perceived as an energy source that is more reliable and less harmful to the environment in the long term than are domestic fossil fuels, which are depletable and increasingly expensive to extract. Oil is a less desirable long-term source of energy for Eastern Europe because of price fluctuations, the USSR's desire to earn hard currency from oil exports, and oil's importance for the petrochemical industries and for other nonfuel applications. Finally, Soviet oil production has peaked and apparently begun its predicted decline. The USSR has 9 percent of total world reserves, but the location of many of these reserves in remote fields entails high extraction costs. In view of these considerations, continuing to supply oil to wasteful East European industries simply does not make sense. Nuclear power will remain a crucial link in the alliance.

The CMEA member countries are currently developing power grids linking nuclear power stations in the western USSR to Eastern Europe. In 1984, the Soviet Union exported 18.8 billion kilowatt hours to the six European CMEA members. Of this, Hungary received 8.8 billion kilowatt hours, or one-third of its electrical energy (all of which came from the Chernobyl plant). Bulgaria received 10 percent of its total electrical energy from the Soviet Union in the same year.[37] In addition to supplying electricity directly, the Soviet Union provides components for East European nuclear plants. The East European countries frequently provide capital for power stations constructed on Soviet territory; and the Czechoslovaks in the past have built water-cooled reactors for the Soviets and are now the major suppliers for Eastern Europe. One Western expert points out that through Soviet–East European cooperation the USSR hopes to prevent its allies from establishing links to capitalist nuclear programs, as well as to enhance its political control over the region and reap economic benefits.[38] The Soviet commitment to expanding nuclear power is not likely to waver, despite a probable slowdown in commissioning new plants in the near future, because nuclear power is generally considered the most cost-effective source of electricity for the European USSR. Soviet concern over continued East European reliance on dirty brown coal and lignite, which adds to the pollution of the most heavily populated Soviet areas, may be another factor encouraging the further expansion of nuclear power.

The Chernobyl disaster has raised fears of nuclear power in Eastern Europe but is not likely to do more than slow plans to expand nuclear power production. Chernobyl brought mixed reactions in Eastern Europe. In Czechoslovakia, for example, a dissident group called "Antiatom" emerged opposed to that country's nuclear power program. More than two thousand Poles demonstrated in Cracow in the aftermath of the accident against the atomic power station being built at Żarnowiec, while a group of engineers protested the experimental nature of the containment structure for the reactor there. In the GDR, petitions were submitted to the government calling for a referendum on nuclear power, while peace groups connected with the Evangelical Lutheran church protested to the Council of Ministers.[39] Hungarian officials, however, suppressed coverage of the disaster and refused to release information on radiation levels in food supplies. The Hungarian ecological group Danube Circle declined to jeopardize its tenuous position by overtly criticizing nuclear power or Chernobyl. In Romania, President Ceauşescu criticized the Soviet handling of the accident but failed to release information to the general public.[40]

Soviet citizens have also demonstrated an acute concern over the issue of nuclear accidents in the wake of Chernobyl. Authorities have been deluged with letters questioning the safety of nuclear power, and newspapers frequently publish interviews with officials and experts designed to reassure the public that adequate safeguards are being adopted. Although sympathetic in their response to popular concerns, Soviet officials steadfastly maintain that they "have no other economically feasible alternative except accelerated development of atomic power."[41]

Two remaining areas of environmental cooperation through CMEA include the coordination of environmental statistics and joint efforts to respond to problems of urban development. In the area of city planning, the CMEA Standing Commission on Construction has worked with Soviet and East European institutes to address problems of air, water, and soil pollution in urban areas, the recultivation of territory, waste utilization, reduction of urban noise, and the setting of basic principles for urban construction.[42] In the 1970s the Standing Commission on Statistics began to develop a methodological base for environmental monitoring, approving a set of indicators and a methodology for their calculation in 1975. A system of 150 indicators was established relating to land and forest use, water and air pollution, excessive noise, industrial wastes and sewage, and capital investment in environmental protection; in 1981 the CMEA members agreed to develop and modernize the system further.[43] A 1984 Soviet study asserted, however, that more work was needed in developing environmental statistics in order to construct an overall information system that would facilitate national planning and further international cooperation.[44] One indication of slow progress in the collection of environmental data is the fact that the Czechoslovak government mandated the central collection and dissemination of such data as recently as mid-1986.[45]

CONCLUSIONS AND OBSERVATIONS

Soviet–East European cooperation on environmental protection reflects the general state of relations among the socialist allies. Environmental agreements are frequently bilateral rather than multilateral. Integration has proceeded slowly in this area, as it has within the CMEA in general. The resource-rich Soviet Union supplies fuel, electricity, and raw materials to the East Europeans in exchange for manufactured products, capital, and political support. The full advantages of a greater division of labor, however, are hampered by a lack of complementarity in production and by a reluctance to grant substantial powers to supranational organizations. Romania's open resistance to increased political and economic integration, motivated by a fear of Soviet domination within CMEA, is the most extreme manifestation of a growing nationalism among the East European countries. In turn, the Soviet leaders appear to be increasingly critical of what they perceive as East European ingratitude for the USSR's willingness to shoulder the economic, energy, and ecological burdens of much of the region. Recent mutterings from Moscow imply that Eastern Europe will be expected to carry more of its own weight in these three areas in the future.

In the past, CMEA environmental protection programs have avoided sensitive political questions, focusing primarily on technical questions and research collaboration. The Soviet obsession with secrecy has in the past influenced the treatment of environmental problems both within the USSR and in bloc relations. Within the CMEA, any criticism of fraternal socialist systems that might imply serious disagreement inside the community has been discouraged. This type of secrecy, which has in the past undermined efforts at environmental protection among CMEA members, may change as a result of Gorbachev's *glasnost*. But we can expect that the sensitive issue of intrabloc relations will be one of the last areas to be opened to public scrutiny.

Developments in Soviet–East European environmental cooperation have paralleled the general course of relations within the CMEA.

Cooperation on narrow technical questions began about the time of Khrushchev's removal, with the exchange of technical information and the initiation of several joint projects on water pollution. After 1971 the emphasis in CMEA was on comprehensive integration through central planning, with concessions to those members unwilling to surrender any more of their national sovereignty. Cooperation on water supply and water pollution, air pollution, and urban development expanded in the 1970s and into the 1980s. Although CMEA cooperation on environmental questions is important to all members, the substantive benefits have in the past been relatively more important to the East Europeans. The Soviet Union's vastly greater territory and immensely larger stores of natural resources distinguish its environmental situation from that of its smaller, more densely populated, and resource-poor East European neighbors. In the reform-conscious 1980s, though, Soviet largesse appears to have reached its limits. Pressures on the East Europeans to make better use of scarce natural resources through conservation and improved efficiency can surely have environmentally beneficial effects. But certain contradictory demands that the East Europeans supply the USSR or other CMEA members with low-cost raw materials (as in the case of Polish coal and metal), or continue to pursue nuclear power development, or expand reliance on heavily polluting brown coal and lignite deposits will more than likely offset the environmental advantages of Gorbachev-style restructuring in the CMEA.

NOTES

The research for this paper was in part funded by a grant from the University of Louisville College of Arts and Sciences. I thank Wlodzimierz Ryzdkowski, Helmut Schreiber, and Sarah Terry for their helpful comments on an earlier draft of the manuscript. Joan DeBardeleben's skillful editing greatly improved the final product.

1. "Gorbachev CPSU Central Committee Political Report," in *Foreign Broadcast Information Service* (*FBIS*, Soviet Union), 25 February 1986, p. 8.
2. In November 1985, the first high-level meeting of the US-USSR Joint Committee on Cooperation in the Field of Environmental Protection since 1979 was held in Moscow. A total of 38 projects (revised from the original 42 agreed to in 1972) was approved one day before the Geneva summit. A follow-up meeting was held in December 1986, at which a protocol on various aspects of air and water pollution, stratospheric ozone depletion, climate change, acid rain, and oceanic pollution was signed. Fitzhugh Green, "The Amerikanskis are Coming," *EPA Journal* 12 (January–February 1986): 21–22; and Fitzhugh Green, "Sparks of Bilateral Congeniality," *EPA Journal* 13 (January–February 1987): 38–39.
3. For discussions of Soviet environmental protection focusing on the Brezhnev era, see Charles E. Ziegler, *Environmental Policy in the USSR* (Amherst: University of Massachusetts Press, 1987); Joan DeBardeleben, *The Environment and Marxism-Leninism: The Soviet and East German Experience* (Boulder: Westview Press, 1985); and Thane Gustafson, *Reform in Soviet Politics: Lessons of Recent Policies on Land and Water* (Cambridge: Cambridge University Press, 1981).

4. CMEA is an organization of states formed in 1949 to facilitate trade and economic development. Active membership currently includes the GDR, Poland, Czechoslovakia, Bulgaria, Romania, Hungary, Mongolia, Cuba, and Vietnam. Yugoslavia is an associate member.

5. R. A. Novikov, ed., *Problema okruzhaiushchei sredy v mirovoi ekonomike i mezhdunarodnykh otnosheniiakh* (Moscow: Mysl', 1976), pp. 224–25. See also B. Gorizontov and V. Prokudin, "Environmental Protection in Comecon Member Nations," *Problems of Economics* 21 (December 1978): 24–40.

6. The Comecon statute on the Council for Protection and Improvement of the Environment is reproduced in *A Source Book on Socialist International Organizations*, William E. Butler, ed. and trans. (Alphen aan den Rijn, The Netherlands: Sijthoff and Noordhoff, 1978), pp. 183–86.

7. Z. Ia. Sheinin, *Ekonomicheskii rost, resursy, i mezhdunarodnoe sotrudnichestvo* (Moscow: Nauka, 1984), pp. 100–102.

8. B. Gorizontov, "The CMEA Countries Solving Ecological Problems," *International Affairs* (Moscow) 6 (June 1980): 113.

9. Josef M. Van Brabant, *Socialist Economic Integration: Aspects of Contemporary Economic Problems in Eastern Europe* (Cambridge: Cambridge University Press, 1980), p. 248.

10. E. M. Goncharova, "Vliianie vodno-ekologicheskikh uslovii na razvitie i razmeshchenie promyshlennosti evropeiskikh stran-chlanov SEV," *Voprosy geografii: geografiia khoziaistva stran-chlenov SEV v usloviiakh integratsii* (Moscow: Mysl', 1984): 300–309.

11. Iosef Vintse, "Vodokhoziaistvennye voprosy basseina r. Tisy," *Ekonomicheskoe sotrudnichestvo stran chlenov SEV* 1 (1981): 31–36.

12. A. Kozyrev, "Sotrudnichestvo v okhrane okruzhaiushchei sredy," *Ekonomicheskie nauki*, no. 3 (1981): 38–39.

13. Vintse, "Vodokhoziaistvennye voprosy basseina r. Tisy," 36. Vintse is vice-chairman of Hungary's State Department of Water Economy.

14. Anton Sikora and Peter Tsetsko, "Vodoinform' na sluzhba prirody," *Ekonomicheskoe sotrudnichestvo stran-chlenov SEV* 8 (1986): 63–65.

15. Miloslav Bogach, "V tseliakh okhrany okruzhaiushchei sredy," *Ekonomicheskoe sotrudnichestvo stran-chlenov SEV* 5 (May 1984): 40–42.

16. Jan S. Prybyla, "The Dawn of Real Communism: Problems of Comecon," *Orbis* 29 (summer 1985): 398.

17. O. A. Chukanova, ed., *Nauchno-tekhnicheskoe sotrudnichestvo stran SEV: spravochnik* (Moscow: Ekonomika, 1986), pp. 221–23.

18. Sheinin, *Ekonomicheskii rost*, p. 102.

19. Oleg Kolbasov, "Pravovoi rezhim pogranichnykh vod," *Ekonomicheskoe sotrudnichestvo stran-chlenov SEV* 4 (1985): 70–73.

20. See Ziegler, *Environmental Policy in the USSR*, pp. 144–47; and Henry Krisch, "Soviet and East European International Environmental Action 1972–85," *Coexistence* 23 (1986): 267–81.

21. John M. Kramer, "The Environmental Crisis in Eastern Europe: The Price for Progress," *Slavic Review* 42 (summer 1983): 218.

22. Kozyrev, "Sotrudnichestvo v okhrane okruzhaiushchei sredy," 39–40.

23. Kramer, "Environmental Crisis in Eastern Europe," 218.

24. The other signatories are Austria, Belgium, Canada, Denmark, Finland, France, the FRG, Italy, Lichtenstein, Luxembourg, Netherlands, Norway, Sweden, and Switzerland.

25. N. H. Highton and M. J. Chadwick, "The Effects of Changing Patterns of Energy Use on Sulfur Emissions and Depositions in Europe," *Ambio* 11 (1982): 324–29.

26. Lester R. Brown et al., *State of the World 1987: A Worldwatch Institute Report on Progress toward a Sustainable Society* (New York: W. W. Norton, 1987), p. 187.

27. Keith Crane, *The Soviet Economic Dilemma of Eastern Europe* (Santa Monica: Rand, 1986), p. 35.

28. Brown et al., *State of the World 1987*, pp. 186–87.

29. Crane, *The Soviet Economic Dilemma of Eastern Europe*, p. 36.

30. N. V. Alisov and A. V. Petrov, "Ekologicheskie problemy razvitiia i razmeshcheniia promyshlennosti evropeiskikh stran-chlenov SEV," *Voprosy geografii: geografiia khoziaistva stran-chlenov SEV v usloviakh integratsii*, vd. 123 (Moscow: Mysl', 1984): 292–300.
31. Ibid.
32. U.S. Congress, House, Subcommittee on Europe and the Middle East, *Poland's Renewal and U.S. Options: A Policy Reconnaissance*, 100th Cong., 1st sess., 5 March 1987, p. 23.
33. I am indebted to Joan DeBardeleben for these figures taken from Jochen Bethkenhagen, "Die Kernenergiepolitik der RGW-Länder," *DIW-Wochenbericht 53*, no. 25/86 (19 June 1986): 309–11.
34. John M. Kramer, "Chernobyl and Eastern Europe," *Problems of Communism* 35 (November–December 1986): 41.
35. Andranik Petros'iants and Aleksandr Panasenkov, "Energiia atoma-sozidaniiu," *Ekonomicheskoe sotrudnichestvo stran-chlenov SEV* 9 (1985): 12–16.
36. *Wall Street Journal*, 4 November 1986, p. 1.
37. Judith Thornton, "Soviet Electric Power after Chernobyl: Economic Consequences and Options," *Soviet Economy* 2 (April–June 1986): 160–61.
38. Kramer, "Chernobyl and Eastern Europe," p. 46.
39. Ibid., pp. 41–43.
40. "Chernobyl and Eastern Europe: One Year after the Accident," *Radio Free Europe Research*, RAD Background Report 67 (24 April 1987); and Frederick Kempe, "Eastern Bloc Is Bound Together by Chernobyl despite Early Forecasts," *Wall Street Journal*, 21 May 1986, pp. 1, 18.
41. Iu. Bublik, "Safety and Power: How Atomic Power Stations Should Develop," *Sovetskaia rossiia*, 30 April 1987, translated in *Current Digest of the Soviet Press* 39 (27 May 1987): 5. One Soviet official expressed it to me more colorfully—"You don't shut down automobile production after one traffic accident."
42. Arnol'd Melikaev, "Statistika okruzhaiushchei sredy," *Ekonomicheskoe sotrudnichestvo stran-chlenov SEV* 6 (1984): 74–75.
43. Kozyrev, "Sotrudnichestvo v okhrane okruzhaiushchei sredy," pp. 40–42.
44. Sheinin, *Ekonomicheskii rost*, p. 103.
45. Katherine Albrecht, "Environmental Policies and Politics in Comtemporary Czechoslovakia" (Paper presented at the Midwest Slavic Conference, Ann Arbor, Michigan, 24 April 1987), p. 14.

4

ENVIRONMENTAL MANAGEMENT IN WEST EUROPEAN COUNTRIES: SOCIAL MOVEMENTS, ECOLOGICAL PROBLEMS, AND INSTITUTIONAL RESPONSES

Philip D. Lowe

Environmental problems are a prominent feature of political life in all West European countries. Over the past twenty years they have posed a mounting challenge to established decision-making systems. Rooted in conflicts outside the sphere of production, environmentalism demands change and adaptation from political structures that are oriented to conventional conflicts between economic interests. But precisely because the environmental movement is not a product of traditional class conflict, the potency of popular environmental concern remains a source of bemusement among political elites, journalists, and academic commentators. There has been a persistent tendency to underestimate the deep social roots of the environmental movement and thus to discount its long-term political potential. Simplistic explanations of its origins have abetted a short-term, reactive approach to pressing environmental problems.

Many social commentators have identified the provenance of the environmental movement in a single event—the publication of a book, the passage of a piece of legislation, the convening of a particular conference, or the occurrence of a disaster. Such explanations are linked implicitly or explicitly with the notion that the environmental movement is a passing fad. Periodically, after a lull in media attention, commentators have been ready to pronounce it a spent force. French politicians, journalists, and academics have been the latest to declare that *les écologistes sont morts*. Invariably, such obituaries have proved premature. Some commentators even suggest that the environmental movement is a manufactured concern or conspiracy: a creation of self-seeking publicists, of media hype and hysteria, of cynical politicians seeking to deflect attention from the *real* issues (whether unemployment, capitalist exploitation, or overseas oppression), or of elitist preservationists seeking to deny the workers their place in the sun.[1] All such assessments, however, fall into the trap of identifying attendant features and effects as fundamental causes.

The starting point for any considered assessment of the environmental movement must be a recognition that it is no flash in the pan, but rather an established feature of all advanced capitalist societies (as well as some socialist societies and many industrializing nations).[2] Environmental concern is a widespread aspect of contemporary West European society, as confirmed by a host of opinion surveys.[3] It is intimately associated with many prominent and progressive social trends—including changing patterns of daily work and living, employment and recreation, and residence and consumption. Perceptions of environmental quality are closely tied up with general notions of social well-being and the quality of life. The protection of the environment has come to be a crucial indicator by which Western publics judge the performance of political and industrial leaders and of governing institutions.

Any explanation for such a complex and ubiquitous phenomenon must be firmly rooted in an analysis of contemporary social and economic change. The environmental movement is a product of secular changes in the social and economic structure of West European, as well as other, societies and, specifically, is an expression of the popular response to the predicaments and benefits that arise from the changing organization and scale of production and consumption.[4] Most commentators date the emergence of the contemporary environmental movement to the late 1960s, but they focus merely on overt political action and manifestation. To understand how at this time there came to be such a ready, attentive public for the efforts of the new breed of environmental activists, we must look to earlier and deeper changes in postwar society.

THE POSTWAR ROOTS OF THE ENVIRONMENTAL MOVEMENT

The immediate postwar years were dominated economically by the reconstruction of manufacturing industries, the revival of agriculture, and the shift from war production to provision for domestic needs. Initially the emphasis was on necessities, but after the drabness and deprivation of war there was also an insatiable demand among European publics for convenience goods and consumer durables. There was unquestioned political commitment to industrial expansion and full employment. The period saw a gradual relaxation of the direct governmental controls over industrial and agricultural production that had been introduced during the war.

Little regard was given to the environmental consequences of industrial expansion: the overriding priority was to rebuild economies and war-ravaged cities. Steps were taken with mixed success in various countries toward comprehensive planning, zoning, and redevelopment of towns and cities that had suffered extensive bomb damage.

Environmental groups did exist from prewar days, often dating back to the turn of the century or earlier. These were elitist, upper-class organizations enjoying little if any popular support, and their concerns were largely with rural or historic preservation.[5] Their limited influence was further compromised in a number of countries, such as France, Germany, and Italy, by the extent to which they or their ideas had been discredited by association with fascist movements during the 1930s and 1940s.

The sustained expansion of West European economies between the late 1940s and the early 1970s was unprecedented in its scale and duration.[6] Not only were the regions of traditional heavy industry affected; the countries of southern Europe, for example, were transformed from rural and agrarian societies into urban and industrialized ones.[7] Inevitably, there was enormous physical upheaval, with massive environmental problems in its wake. The popular response to these problems was mediated by the changed social outlook and structure of the consumer and leisure culture that emerged in the 1950s and 1960s.

Sustained economic growth provided greatly increased discretionary incomes, which opened up mass markets for such items as automobiles and television sets. New and improved standards of housing became both personal and public priorities. Growth in labor productivity also meant shorter working hours, which resulted in a shorter work week, more holidays, and longer time spent by teenagers in full-time education and training.

Each of these different trends was linked with changes in values that gave rise to new constituencies for environmental politics. The greater opportunities for leisure led to increased interest and participation in outdoor recreation, which made people sensitive to the value of, and the threats faced from, open spaces and natural environments.[8] The shift in emphasis in daily living from a preoccupation with work to a greater role for home and leisure also focused attention on the residential environment and its amenities. These new priorities were especially marked for those mainly middle-class families who, through the automobile and access to private housing markets, were able to distance home from work by moving to the suburbs or beyond.[9]

A somewhat different constituency for environmental politics was furnished by the burgeoning student population. People who are better educated are likely to have greater knowledge and understanding of the environment and to be more sensitive to the threats it faces. Extensive surveys have shown that proenvironmental attitudes are positively correlated with levels of education.[10]

The expansion of higher education, and of government and industrial research institutions, led to a huge growth in the scientific community. The physical sciences and the more reductive and experimental biological sciences benefited most through their perceived relevance to industrial, defense, agricultural, and medical needs. Practitioners of the softer natural sciences found themselves becoming increasingly marginal. Many of them, in looking for ways to reassert their status and redefine their social relevance, were attracted to investigations of the environmental consequences of urban-industrial expansion, and especially of the new science-based technologies. Ecological studies began to expand considerably, particularly in the 1960s, and some of the practitioners sought a wider audience for their results.[11]

Finally, the role of the mass media should be mentioned in creating a latent although attentive public for environmental issues. The contemporary environmental movement has grown up with television. This medium established an appreciative mass audience for the televisual aspects of the environment, especially natural history and the countryside. Television also has brought into people's living rooms vivid and often dramatic pictures of pollution accidents and environmental disasters near and far. Perhaps more fundamentally, it has informed people about the qualities of different places and localities and has enabled them to make knowledgeable comparisons with their own environmental well-being.

One of the significant features of modern social movements like the environmental movement is that the media rather than the organizational leaders convey the message to the mass public. This feature contrasts with major social movements of the nineteenth century, such as the labor movement, in which the leaders unquestionably had to be charismatic figures. The environmental movement, instead, has been able to draw upon the potential of the media to inform and manipulate mass opinion. Unlike nineteenth-century movements, for example, there is no longer a necessary reliance on organic solidarity, group consciousness, and strong centralized organization to maintain cohesion. The environmental movement, in contrast, is only loosely organized and coordinated, and it comprises diverse interests. Indeed, without the mass media to focus its frequently diffuse sentiments on some common purpose, it would be difficult to regard such a loosely structured grouping as a social movement at all.[12]

Environmental politics, sporadic and marginal in the 1950s and early 1960s, gradually began gathering momentum in the late 1960s as a result of these social changes. Several countries during this period adopted legislative measures, often involving the extension and strengthening of public health and public nuisance legislation dating from the turn of the century or earlier, to control the grossest aspects of air and water pollution and waste disposal that threatened the viability of continued urban and industrial expansion.[13]

But some of the newer and expanding sectors, such as the chemical and electrical industries, were creating new environmental hazards as well as problems of toxic waste disposal. New sources of energy brought new problems too: radioactive waste and the marine and coastal pollution caused by recurrent oil spills. Moreover, emerging patterns of consumption were very wasteful. As litter and the casual dumping of defunct "durables" mounted, so did the problem of disposing of nondegradable domestic refuse. The intensity of consumption also generated pollution problems, ranging from automobile emissions to synthetic detergents.

THE EMERGENCE OF THE ENVIRONMENTAL MOVEMENT

The late 1960s and early 1970s saw a continued expansion of the consumer society. The number of cars in Western Europe doubled during this period. Suburban growth continued apace, and the phenomenon of counterurbanization emerged as rural areas came under pressure from people seeking commuter, second, or retirement homes.[14] Increased mobility brought its own problems, such as congestion, urban sprawl, loss of open countryside and coast, noise, and recreational conflicts. Indeed it became apparent that as more and more people acquired the means of access to attractive residential and recreational environments, this access threatened the very attractions that they sought. The growing awareness that environmental amenity was a scarce *positional good* (in other words, a commodity subject to social scarcity) contributed to an emerging consciousness of the social limits to growth, expressed through the proliferation of local protectionist associations and campaigns.[15]

Conflicts over land use planning and local resistance to development were given an added twist from a different direction. Continued economic growth and restructuring in the face of declining profitability demanded a growing role for government as the promoter of major infrastructural projects (such as the construction of highways, airports, reservoirs, power stations, and new towns), major schemes of land drainage, and the redevelopment of city centers. These projects were often bitterly opposed by local residents, businessmen, and tradespeople. The role of government as both initiator and arbiter of these developments gave local opposition a strong political dimension.

Out of this heightened concern for amenity emerged the modern environmental movement. Broadened popular appeal revived the fortunes of existing preservationist groups, which now experienced massive growth in membership. Many new environmental groups were formed. In Britain alone, at least twenty-three national environmental groups and seven hundred local amenity societies were formed between 1966 and 1975.[16] The Commission of the European Community (European Commission) in 1975 identified more than twenty thousand associations for the protection of the environment in countries belonging to the European Community and considered this estimate low.[17]

A major contributing factor was heightened media interest fueled by various national and international campaigns, such as the European Conservation Year (1970) and the United Nations Conference on the Human Environment (1972). Public concern was also aroused by a series of alarming environmental disasters—Torrey Canyon, Santa Barbara, Minimata, fish kills in the Rhine, and the Irish seabird crash. Amid a gathering sense of impending environmental crisis, with a number of pundits predicting imminent ecological collapse, many of the new environmental groups questioned the direction of society. They tended to be more radical in their aims than the established preservationist groups and to take a holistic view of environmental problems.[18] Many of the activists had belonged to the student movement of the late 1960s, and some of them saw environmental issues as a challenge to the dominant economic and political order of capitalist societies. The profligate use of nonrenewable natural resources was identified as the most serious problem, and economic and population growth as the fundamental cause. Many people accustomed in their daily lives to the experience of social limits to growth (noise, congestion, urban sprawl) readily embraced the notion of physical and ecological limits to growth (as set by the absolute availability of natural resources and the capacity of ecological and geophysical systems to absorb urban and industrial wastes).[19]

Widely publicized investigations of the effects of two particular types of pollutant—radioactive fallout from nuclear tests and the residues of persistent organochlorine insecticides—also caused profound shifts in popular attitudes.[20] Alarming evidence of the universal distribution of these toxic substances (DDT was even found in 1965 in penguins in the Antarctic) and their concentration in food chains served to symbolize powerfully the potential dangers of global environmental pollution. Perceptions of a radioactive and chemical universe—invisible, all-pervasive, and insidious—were profoundly disturbing. They not only blighted faith in scientific and technological progress but also blocked the former escape route of geographical mobility whereby the middle classes had distanced themselves from the older and more local environmental problems arising from slums, smoke, and sewage. Increasingly, local protectionist interests identified with wider national and international campaigns for environmental protection.

European governments responded to these environmental pressures by adopting new consultative procedures and forms of analysis that incorporated environmental considerations. Public participation was promoted in planning decisions.[21] Proposals for particularly contentious developments were subjected to referenda or public inquiries, or environmental impact analyses.[22] In almost all West European countries, general and specialized environmental laws were passed and institutions set up to implement them.[23] In some countries, the environmental responsibilities of existing ministries were enlarged. This was the case, for instance, with the ministries of agriculture in Sweden, health in the Netherlands, and interior in Germany and Finland. In other countries such as the United Kingdom and Spain, large superministries responsible for the environment, public works, and urban affairs were formed. In yet other countries, such as Norway and Denmark, more specialized ministries of the environment were created. Environmental objectives were defined, including quality or ambient standards for air and water, and also goals for the conservation of natural resources and the enhancement of amenities. Pollution controls, particularly on new industries, were made much more stringent. Public investment was greatly increased in the creation of parks, nature reserves, and recreational areas, and in waste disposal and treatment facilities.[24]

THE TESTING OF THE ENVIRONMENTAL MOVEMENT

The fourfold increase in oil prices of 1973 and its impact on European economies had major consequences for environmental politics. The faltering growth rates, inflation, recession, and mass unemployment experienced since then have polarized European societies and undermined the political consensus on which the key postwar policies for social and economic management had been built.[25] Environmental policy did not remain unscathed.

The most general consequence was a slowdown in public and private investment in pollution control and environmental improvements. Pressure mounted from industrialists for relief from environmental controls, and governments showed growing reluctance to press cleanup measures on firms hit by recession, especially in areas of high unemployment. With governments responding to inflation by cutting back on public expenditures, much needed capital investment in waste treatment facilities was either shelved or indefinitely postponed. Private firms were also less willing or able to replace outdated "unclean" plants or to invest in the latest equipment for pollution control. The large-scale closure of plants (especially in traditional heavy industries), however, eliminated many long-standing sources of air and water pollution but also left a legacy of extensive dereliction and blight in depressed industrial regions.

One welcome environmental consequence of the oil shock was the impetus and urgency it gave to the conservation of resources. The consequent drive to save energy, often with governmental backing, also had positive implications for the efficient use of other resources, either because energy saving or recovery facilitated other forms of recycling, or because increasingly cost-conscious manufacturers were induced to look for other ways to reduce their raw-material costs. In turn, these efforts alleviated problems of pollution and waste disposal. Another positive development was the extent to which conservation and environmental improvement projects were adopted in government-sponsored programs to create jobs in countries such as Germany, Sweden, Denmark, France, and the Netherlands.[26]

The implementation and elaboration of environmental policies have encountered not only a worsening economic climate and the political resurgence of business interests alarmed by the "limits to growth" debate, but also the emergence of "new right" politicians unsympathetic toward business regulation and *dirigiste* (interventionist) policies of any kind. By and large, these counterpressures have not led to the wholesale dismantling of environmental controls, but in certain cases they have forced cuts in budget and staff of regulatory agencies and, in others, won special exemptions and relaxations. There have also been curbs on the resources for environmental research and monitoring, and efforts to circumvent or place bounds on participatory and consultative procedures.[27]

In the main, the institutionalization of environmental reforms has been sufficiently secure to withstand such assaults. The emphasis, though, has been increasing on defending and implementing past measures rather than seeking new breakthroughs. This has been done by consolidating the position of environmental agencies and related professional networks of planners, technicians, and researchers. Efforts have been concentrated on rationalizing environmental management, improving the efficiency and precision of regulatory measures, and integrating environmental policy objectives with other policy fields.[28]

Public support for environmental conservation has remained buoyant.[29] From opinion surveys it is evident that only a small minority of the West European public is willing to sacrifice some degree of environmental improvement for economic growth; indeed, a sizable majority gives priority to protecting the environment even if doing so means restricting economic growth.

One important factor underpinning this continued popular support has been a marked change in the employment structure. The focus for new employment since the 1960s has been increasing in the service sector, which in almost all West European countries (except Greece, Portugal, and Spain) now provides most of the jobs. Thus a growing majority no longer owe their livelihoods to forms of employment (whether in agriculture, mining, construction, or manufacturing) that involve the manipulation or processing of natural resources and that inevitably pose the most acute environmental problems. Fewer and fewer people, therefore, are exposed directly—through their work and income dependency—to a personal conflict between economic and environmental welfare. Opinion surveys confirm that environmental concern tends to be most strongly expressed by those employed in the burgeoning service sector.[30]

With their popular support undiminished, the response of environmental leaders and activists to the slowdown of environmental reforms has diverged. Some have been drawn into consultative arrangements and the monitoring of environmental measures, and these have increasingly emphasized their technical contribution to the regulatory process. Other activists have reacted to the harsher economic and political climate with more radical tactics, including demonstrations, civil disobedience, nonviolent direct action, and, in isolated cases, sabotage. Those adopting a more adversarial stance have looked beyond consultative arrangements and tactics of influence toward strategies of confrontation and power.[31]

One issue stands out for its potential to radicalize and politicize environmentalists—nuclear power. After the oil shock of 1973, most West European governments launched crash programs to develop nuclear energy. The political priority attached to the construction of nuclear plants and the vehement opposition they provoked strained public consultation procedures to the breaking point. The antinuclear movement became a rallying cause for oppositional elements, including radical environmentalists, new-left activists, and disaffected local populations at the threatened sites.[32]

The coalescing of these oppositional elements around certain electoral strategies has led to the formation and growth of Green parties. Starting with the Ecology Party in Britain in 1974, such parties have emerged in most West European countries over the last few years: only Norway and Greece do not have any.[33] In Austria, Belgium, Finland, Italy, Sweden, Switzerland, Luxembourg, and West Germany, these new parties are represented in the national parliaments. The German and Belgian Greens and a Dutch Green list won seats in the 1984 elections to the European Parliament. These developments have forcefully reinstated environmental issues at the top of the political agenda, especially in countries such as West Germany, Sweden, and Belgium, where Green politicians have assumed office in regional and local government. In other countries, including Norway, Denmark, and Britain, the established political parties, aware of the potential of the Green vote, have strenuously competed in presenting their environmental credentials to the electorate.

THE COMPARATIVE POSITIONS OF NATIONAL
ENVIRONMENTAL MOVEMENTS[34]

The picture presented here of environmental trends in Western Europe is one of rising environmental problems in the postwar years, of public reaction to these in the context of a growing interest in environmental conservation, and of government response to expressed concerns.

At a very general level one can point to key social variables that broadly facilitate popular concern and its political expression. By and large the social priority given to environmental issues rises with the maturing of a country's economic development. I am referring here not only to a nation's relative wealth, but also to its history of urbanization and industrialization and the balance of its employment structure between the primary (extractive and agricultural), secondary (manufacturing), and tertiary (trade and services) sectors. Thus the institutionalization of environmental concern tends to be strongest and of longest standing in the mature, postindustrial economies of northern Europe, and weakest and least developed in the economically less advanced countries of the European periphery, such as Greece, Portugal, and Ireland.

Nowadays, however, popular concern for the environment is widespread across all these countries. It seems to be a product of two separate factors: first, sensitivity toward and awareness of pressures on the environment; and second, direct experience of environmental deterioration or the threat of it. The first tends to be stronger among the better educated, those with higher incomes, and those employed in the tertiary sector. These people, however, are also more likely to have the personal resources and information to be able to avoid the second factor. Moreover, the countries experiencing the highest incidence of deteriorating environments—those of southern Europe that are rapidly urbanizing and industrializing but have relatively weak regulatory structures—have (on average) lower incomes, lower levels of education, and lower levels of employment in the tertiary sector.

Although popular environmental concern is a common feature of all West European countries, the nature of that concern varies among social groups and countries according to the relative incidence of the two factors. People with lower incomes and less education, for example, tend to place greater emphasis on everyday problems with an immediate impact, such as litter, noise, and vehicular fumes—issues that are also given higher priority in the public opinion of Portugal,

Spain, Italy, Greece, and Ireland than elsewhere. In contrast, people with higher incomes and more education tend to emphasize more the problems with less immediate and tangible impact, such as acid rain, industrial pollution, pesticides, and wastage of natural resources— issues that are given greater emphasis in the public opinion of Germany, Denmark, the Netherlands, and Luxembourg. Opinion in France, Belgium, and Britain occupies a median position between these poles.[35]

The divide between the more advanced and less advanced economies is also reflected in the balance of concern between local and nonlocal issues. *In all countries* there is greater concern over national and global environmental problems (marine and river pollution, industrial waste disposal, and loss of species) than local problems (noise, landscape damage, and the quality of air and drinking water). Regarding nonlocal issues, there is not much transnational variation, and the degree of concern is strongly correlated with levels of income and education. Regarding local environmental issues, however, there is a marked variation, and concern is much *less* strongly income and education dependent. The publics of Spain, Portugal, Italy, and Greece, in particular, are much more likely to harbor complaints about their own local environmental conditions than are the publics of northern Europe, and this likelihood seems to be related to the pressures of rapid urbanization.[36] On the whole, expressed dissatisfaction with the local environment increases with population density: city dwellers express, on average, twice the level of complaint as country dwellers; and residents of apartment blocks, twice the level of residents of detached houses. For Spain and Italy, at least, this local concern is matched by strong concern with national and global issues. This pattern contrasts with Denmark and the Netherlands in particular, where publics exhibit very low levels of dissatisfaction with the local environment but very strong concern over nonlocal problems. Greece presents just the opposite pattern, whereas Ireland registers relatively low on both accounts. These different patterns can be explained by the differential incidence of aware and sensitized publics on the one hand and deteriorating environments on the other.

The next question is how this popular concern is translated into action. A 1986 survey conducted by the European Commission constructed an index of willingness to take action to protect the environment, which included such behavior as recycling waste, refraining from littering, converting car exhausts, becoming involved in local action, taking part in a demonstration, and supporting an environmental group. Willingness was found to depend on the respondent's degree of concern about the national environment and to increase with the level of education and income. These findings suggest that willingness to act may be a function of ability to mobilize resources, that is, of access to relevant private, political, and informational means. This hypothesis is reinforced by a comparison between countries: the average citizen is least inclined to take action in Ireland, Greece, and Portugal, and most inclined in Luxembourg and the Netherlands. Indeed, there seems to be a straightforward relationship between national prosperity and public willingness to take action to protect the environment.[37]

The opportunities for expressing environmental concern politically depend also on the autonomy and diversity of civil society. Key factors include the degree of openness of government, the acceptance of the legitimacy of interest groups, state and societal tolerance of dissent, the diversity and freedom of the mass media, and the plurality and independence of the scientific community. Thus the countries of northwestern Europe, with their long traditions of liberal democracy, have much stronger and better organized environmental lobbies than either France, with its *dirigiste* state, or such countries as Spain, Portugal, and Greece, which have only recently emerged from authoritarian rule. The European Commission's survey found that public involvement in an association to protect the environment was highest in Denmark, Luxembourg, the Netherlands, Italy, and the United Kingdom, and lowest in Spain, Portugal, Germany, and Greece. Civic concern with the environment in the latter group of countries, according to the survey, finds its outlet more in protest demonstrations and (especially in Germany) local action to improve the environment. A strong environmental lobby, it should be noted, is certainly not a necessary condition for the introduction of environmental legislation, but it may be crucial in challenging entrenched economic or state interests opposed to environmental reforms and to ensuring that legislation, once passed, is actually implemented.

A final systemic factor relates to the traditions and procedures for resolving social conflicts and accommodating popular demands. In very general terms, one can distinguish two contrasting approaches. The first is through bureaucratic accommodation, participatory structures, and the incorporation of relevant interests into consultative procedures. The second is by formal and constitutional means, including adjudication, plebiscite, parliamentary resolution, and the electoral process. These two approaches are not mutually exclusive. All West European countries, of course, possess a variety of constitutional mechanisms for resolving social conflicts; conversely, all West European governments have moved steadily over the past twenty years to accommodate an expanding array of interest groups through bureaucratic negotiations and bargaining. The relative balance between the two does vary, however, between countries. In Britain and Sweden, for example, there appears to be a natural inclination for policymakers (be they politicians, civil servants, or interest group leaders) to "process" problems by consultation with interest groups.[38] Indeed, in Britain, resorting to formal, constitutional means to resolve conflicts is comparatively rare and strongly discouraged. Such means still play a significant role elsewhere, especially in countries with a tradition of Roman law and/or a federal constitution, such as Spain, Belgium, West Germany, and Switzerland. Whereas bureaucratic accommodation fosters consensus, compromise, and a problem-solving approach—and usually does so deliberately—constitutional resolution of issues tends to be associated with more conflictual and adversarial politics. In part, this difference lies behind the contrast between the highly integrated, insider lobbying of environmental groups in Britain and Sweden and the more confrontational Green politics of West Germany, Italy, France, and Spain. The contrast also applies within different countries to different issues. For example, an issue such as nuclear energy, which arouses tremendous contention and popular hostility, is apt to provoke constitutional challenges such as litigation, referenda, and electoral opposition in most countries. On the other hand, a more tractable issue such as nature conservation, on which opinion is not sharply polarized, is the subject of conventional lobbying and negotiation in most countries. In many respects, therefore, the dynamics of environmental conflict depend as much on the nature of the issue as on national distinctions of social and political culture.

ASSESSMENT AND CONCLUSIONS

Most countries now have fairly sophisticated policies and institutions to regulate or alleviate many of the more prevalent and most damaging environmental threats. Progress varies by problem and by country: the countries of southern Europe got a relatively late start in their efforts to protect the environment and still lag behind their northern neighbors. Some of the key achievements of policy over the past twenty years include the following:[39]

- reduced air pollution by sulfur dioxide, carbon monoxide, and particulate matter in urban areas;

- improved water quality, particularly in terms of reduced suspended solids and oxidizable organic matter;
- reduced releases to the environment of some persistent chemicals, including mercury, DDT, and PCBs; and
- improved protection for threatened species and habitats.

A number of critical shortcomings remain, however, and new difficulties and hazards loom ahead. The key issues are discussed in the following paragraphs.

The Implementation Gap

If the late 1960s and early 1970s were the key period for environmental policy-making and legislation, then one of the major preoccupations of the 1980s has proved to be the implementation of the relevant laws and policies. The growing literature on the impact of environmental controls and regulations identifies a considerable "implementation deficit" between legislative intent and policy results.[40] The following causes have been adduced: inadequate understanding of the problems to be tackled, difficulties in devising effective and cost-efficient policy instruments, lack of resources devoted to implementation, lack of trained personnel to administer regulations, conflicts with other policy objectives, poor or nonexistent monitoring, ineffectual penalties and enforcement procedures, and extensive noncompliance. Industrial interests, through their economic and political power and close contacts with regulatory officials, have been able to frustrate, delay, or otherwise contain the implementation process. In general, it can be said that regulatory bodies seldom have the powers and the resources to match the problems and interests they confront.

Second-Order Problems

Initial efforts to reduce pollution have been relatively easy. They have concentrated on the coarsest and most acute, concentrated, visible, and tractable forms of pollution and environmental hazards. Governmental action, for example, has often removed only the most prominent and controversial pollutants while leaving untouched other, perhaps far more dangerous, hazards and so the underlying causes of the problems.

From a comparative research project, Jänicke concluded that efforts to combat immediately apparent pollutants such as dust and sulfur dioxide have been relatively effective but that the resulting "blue skies" had defused much of the concern over air pollution that would have to be mobilized to obtain more than symbolic responses in controlling less visible pollutants.[41] The degree of concentration of some pollutants at "black spots" and of emissions from point sources (such as chimneys and sewage outfalls) has often been reduced or stabilized, but the dispersal of pollutants and emissions from diffused sources has generally increased. In addition, although acute ill health and death caused by short-term exposure to intense loads of pollution have generally decreased, risks to humans of long-term effects from exposure to substances that may give rise to genetic changes, cancer, and birth deformities still demand attention.

Knock-on Effects

In some instances, the expediency of existing controls has generated new and more intractable environmental problems. To a major extent, for example, environmental controls have had an influence on the location of industrial plants, as industrialists have sought to avoid certain restrictions. This avoidance seems to have been a factor in shifting the burden of water pollution from upriver sites to estuarial and coastal sites; now there is growing concern over the deteriorating condition of regional seas such as the North Sea, the Baltic, and the Mediterranean.[42] Peripheral siting of hazardous polluting plants has also increased the incidence of transfrontier pollution. Finally, efforts to clean up "black spots" have too often relied on dispersal (for example, through building taller chimney stacks) rather than on eliminating pollutants. Thus, while local concentrations of sulfur dioxide, nitrogen oxide, and hydrocarbons have declined or stabilized in urban areas, large-scale pollution resulting from them and the long-range transport of their secondary products such as photochemical oxidants and acidic compounds have increased.[43]

Diffuse Sources

Perhaps the most intractable pollution problems arise from diffuse sources, which have multiplied through the following causes: the increasing pollution load from consumption, the growing number and availability of hazardous chemicals, the growing number of motor vehicles, and the growing use of synthetic fertilizers and pesticides by farmers. Trends toward industrial decentralization and counterurbanization are also likely to create new difficulties of pollution control. These are some of the pressing problems:

- pollution of surface water and groundwater, particularly by nitrogenous fertilizers and pesticides, and the consequent eutrophication of lakes and contamination of drinking water;
- the transport, storage, use, and disposal of hazardous chemicals; and
- the pervasiveness and clamor of noise (large numbers of people are now exposed to levels that interfere with their daily life).

International Problems

Environmental issues do not respect national boundaries. Pollutants are carried across borders, and the global commons (the atmosphere, the seas, and international rivers and lakes) are universally abused. Progress in tackling national pollution has focused attention on these transnational and international problems.[44] Solutions are extremely difficult, however, involving sustained, cooperative efforts between states with divergent interests and different political and administrative structures. An additional factor is the potential of national environmental regulations as trade barriers. This factor has stimulated business and governmental efforts to coordinate environmental standards to maintain international competitiveness and has been a spur to international agreement.

The late 1960s and the 1970s saw the signing of a large number of multilateral conventions for the protection of the environment, but progress in implementing them has been very slow, facing as it does three gaps in implementation: the gap between signing a convention and ratifying it, the gap between ratification and the necessary national legislation, and the gap between legislation and what actually happens. For example, problems with financing and disagreements over standards have limited any improvements from the 1976 Barcelona Agreement to control land-based pollution entering the Mediterranean.

The efforts made by the countries of the European Community to coordinate their environmental policies present an interesting exception. The European Community differs from all other international organizations in having powers and institutions of its own that bridge the first implementation gap (community directives are binding on member states) and that can also exert pressure to close the other two gaps.[45] The European Community has the duty to monitor the adoption of its directives in national laws and procedures and the power to ensure that they are applied; if it is not satisfied, it can take a member state before the European Court of Justice. The community's

Fourth Environmental Action Program (1987–92) gives high priority to following up the implementation of community legislation.

The First Environmental Action Program was initiated in 1974. Since then, the European Community has adopted more than a hundred legal instruments covering water quality, air pollution, hazardous chemicals, noise, wastes, protection of wildlife, and the safety of nuclear installations. Although the practical progress that has been made is still modest, the European Community has acquired very extensive and unique experience with the vagaries of transnational cooperation on environmental protection.[46]

New Technologies

The growing technology-based industries such as microelectronics and biotechnology are creating new pollution problems. Toxic materials are used in their production, and cases have been reported of soil and groundwater contamination and other environmental damage arising from leakages of these materials. There is also a lack of knowledge of the environmental pressures that biotechnologies and new chemicals may create.[47] It is estimated that there are eighty thousand organic and inorganic chemicals in commercial production, with between one and two thousand more new chemicals appearing on the market each year. Procedures to regulate their use and impact are woefully inadequate.

Beyond the Piecemeal Approach

Environmental policy-making has tended to be reactive and has failed to promote comprehensive solutions; most governments have preferred a muddling-through approach. Such an approach faces two potential pitfalls. Uncoordinated reactions to different problems and political pressures could result in contradictory, costly, and possibly dysfunctional outcomes. The muddling-through course is also politically vulnerable, for environmental problems are left to evolve in an uncontrolled fashion and new issues and sources of public protest are liable to emerge at any time and in an unpredictable manner. It remains an open question whether a piecemeal approach can successfully contain both the ecological *and* political problems or whether a more comprehensive and anticipatory strategy, such as a "preventive environmental policy," is required.[48]

A few countries have begun to explore such a strategy. One of the approaches being developed includes integrated (or cross-media) pollution controls. This approach recognizes the futility of considering each environmental medium (air, water, and land) in isolation. Examples of this approach include the integrated issuing of permits for industrial plants in the Netherlands, West Germany, and Sweden. West German environmental policy has also begun to elaborate the concept of *Vorsorgeprinzip* (the principle of precaution), whereas Dutch policy consciously seeks to anticipate major environmental problems and to pursue integrated prevention and management strategies.[49]

Legal developments in the European Community are obliging other European countries to take at least some steps in this direction. The so-called Seveso Directive, for example, adopted under pressure from the European Parliament following the explosion in Italy in 1976, places a general obligation on manufacturers to prevent major accidents and requires those using any of about 180 dangerous substances to produce safety reports and emergency plans as a matter of public information. Although several member states have still not introduced the legislation to fulfill the obligations of the directive, it has evidently stimulated manufacturers and authorities to concentrate more on preventive safety matters.[50] Perhaps the most important environmental initiative by the European Community, which came into effect in 1988, is the requirement that the environmental effects of major development projects be formally assessed.[51]

The community's Fourth Environmental Action Program envisages further progress, including the promotion of preventive and cross-media approaches to pollution control. None too soon, therefore, are West European political and business leaders being obliged to give serious consideration to the complex environmental problems being created by their countries' economies.

NOTES

I would like to thank Joan DeBardeleben, David Baldock, Nigel Haigh, and Jaap Frouws for their comments on earlier drafts of this paper.

1. Such views are most often expressed by technocrats, industrial, business and farming leaders, and politicians, mainly of the traditional left or laissez-faire right. One of the earliest and most thoroughgoing attacks on environmentalists was by the then editor of *Nature*, John Maddox, in his book, *The Doomsday Syndrome* (New York: McGraw-Hill, 1972).

2. As yet, little work has been done comparing environmental movements beyond gathering basic information on national organizations. See, for example, B. G. Andersson and S. H. Huveröd, eds., *Kampen för miljön: En lägesrapport om miljörörelser* (Göteborg, Sweden: Centrum för Tvärvetenskap, Göteborgs Universitet, 1979); E. Tellegen, *Milieubeweging* (Utrecht: Aula/Het Spectrum, 1983); L. Lemkov and F. Buttel, *Los Movimentos Ecologistas* (Madrid: Editorial Mezquita, 1983). The U.S. Sierra Club publishes a *World Directory of Environmental Organizations*. The best source of information on environmental movements in developing countries is the

Nairobi-based Environment Liaison Center and its various publications. See also Netherlands IUCN Committee, *Environmental NGOs in Developing Countries* (The Hague: Staatsdrukkery, 1985). On environmental politics in socialist countries, see the other chapters in this volume; also F. Singleton, ed., *Environmental Problems in the Soviet Union and Eastern Europe* (London: Lynne Rienner, 1987); and G. Enyedi, A. J. Gijswijt, and B. Rhodes, *Environmental Policies in East and West* (London: Taylor Graham, 1988).

3. *The Europeans and Their Environment* (Brussels: Commission of the European Communities, 1983 and 1986); *Compendium of Environmental Data* (Paris: OECD, 1985 and 1987).

4. I would contrast this type of explanation, which emphasizes the social response to the changing organization of production, with the social psychological explanation of contemporary social movements promoted primarily by Ronald Inglehart, which emphasizes a general value shift in postwar generations. See R. Inglehart, *The Silent Revolution: Changing Values and Political Styles among Western Publics* (Princeton: Princeton University Press, 1977); and for a critique of this analysis, P. D. Lowe, and W. Rüdig, "Political Ecology and the Social Sciences," *British Journal of Political Science* 16 (1986): 513–50.

5. On the historical roots of environmentalism, for France, see C. M. Vadrot, *L'écologie, histoire d'une subversion* (Paris: Syros, 1978). For Germany, see K.-G. Wey, *Umweltpolitik in Deutschland: Kurze Geschichte des Umweltschutzes in Deutschland seit 1900* (Opladen: Westdeutscher Verlag, 1982); C. Conti, *Abschied vom Bürgertum: Alternative Bewegungen in Deutschland von 1890 bis heute* (Reinbek: Rowohlt, 1984); and K. E. Rothschuh, *Naturheilbewegung, Reformbewegung, Alternativbewegung* (Darmstadt: Wissenschaftliche Buchgesellschaft, 1983). For Britain, see J. Sheail, *Nature in Trust* (Glasgow: Blackie, 1976); and P. D. Lowe, "The Rural Idyll Defended: From Preservation to Conservation," in *The Rural Idyll*, G. E. Mingay, ed. (London: Routledge and Kegan Paul, 1989). For Sweden, see D. Haraldsson, *Skydda vär natur: Svenska Naturskyddsföreningens framväxt och tidiga utveckling* (Lund: Lund University Press, 1987).

6. D. H. Aldcroft, *The European Economy 1914–1970* (London: Croom Helm, 1978).

7. A. Williams, ed., *Southern France Transformed* (London: Harper & Row, 1984).

8. I. G. Simmons, *Rural Recreation in the Industrial World* (London: Edward Arnold, 1975).

9. J. H. Johnson, ed., *Suburban Growth: Geographical Processes at the Edge of the Western City* (London: Wiley, 1974).

10. Lowe and Rüdig, "Political Ecology and the Social Sciences."

11. E. J. Kormondy and J. F. McCormick, *Handbook of Contemporary Developments in World Ecology* (Westport, Conn.: Greenwood Press, 1981).

12. P. D. Lowe and D. Morrison, "Bad News or Good News: Environmental Politics and the Mass Media," *Sociological Review* 32 (1984): 75–90.

13. A historical compilation of environmental legislation in the countries of the European Community is given in *European Environmental Yearbook* (London: DOCTER International UK, 1987).

14. A. J. Fielding, "Counterurbanization in Western Europe," *Progress in Planning* 17 (1982): 1–52.

15. F. Hirsch, *Social Limits to Growth* (London: Routledge, 1977).

16. P. D. Lowe and J. Goyder, *Environmental Groups in Politics* (London: George Allen and Unwin, 1983).

17. European Commission, "Non-governmental Organizations for the Protection of the Environment in the European Community," *Industry and Society*, no. 21/75 (27 May 1975): 3–12.

18. M. Allaby, *The Eco-Activists* (London: Charles Knight, 1971).

19. Inglehart, *Silent Revolution*; S. Cotgrove, *Catastrophe or Cornucopia* (Chichester: Wiley, 1982).

20. See, for example, B. Commoner, *Science and Survival* (London: Ballantine, 1966).

21. D. H. McKay, *Planning and Politics in Western Europe* (London: Macmillan, 1982); and R. H. Williams, ed., *Planning Europe: Urban and Regional Planning in the EEC* (London: George Allen and Unwin, 1984).
22. *Environmental Impact Assessment* (Paris: OECD, 1979).
23. *The State of the Environment in OECD Member Countries* (Paris: OECD, 1979).
24. Ibid.
25. R. Dahrendorf, *European Economies in Crisis* (London: Weidenfeld and Nicolson, 1982).
26. *OECD and the Environment* (Paris: OECD, 1979).
27. A few countries actually downgraded their environmental agencies. The French Ministry of Environment, for example, was subjected to a succession of reorganizations that left it weak and marginal. The post of minister of environment was abolished by the Italian government not long after its establishment in 1973, only to be resurrected in 1983. The British government in 1979–80 scrapped a series of advisory committees and bodies on environmental issues. The OECD has gathered statistics on the amount spent annually by various West European governments on environmental research and development. By and large, the sums peaked between 1977 and 1979 at 1 percent to 3 percent of total government expenditure on research and development, and tailed off subsequently (OECD *Compendium of Environmental Data* 1985, table 14.3). Political concern at the possible restraining effects of environmental controls on economic performance led to macroeconomic analyses of their impact in Finland, France, the Netherlands, and Norway, but these revealed the net macroeconomic effects to be very small indeed; see OECD, *Environment and Economics*, Background Papers, vol. 1 (Paris: OECD, 1984).
28. *The State of the Environment* (Paris: OECD, 1985).
29. *The Europeans and their Environment*, 1983 and 1986.
30. Lowe and Rüdig, "Political Ecology and the Social Sciences."
31. J. P. Pilat, *Ecological Politics: The Rise of the Green Movement* (Beverly Hills: Sage, 1980); L. W. Milbrath, *Environmentalists: Vanguard for a New Society* (Albany: SUNY Press, 1984): Cotgrove, *Catastrophe or Cornucopia*.
32. W. Rüdig, "A Comparison of Anti-nuclear Movements in the United States, Britain, France, and West Germany," in *Public Acceptance of New Technology*, R. Williams and S. Mills, eds. (London: Croom Helm, 1986).
33. Lowe and Rüdig, "Political Ecology and the Social Sciences."
34. This section draws on comparative opinion surveys collated by the OECD and the European Commission; see note 3.
35. *The Europeans and Their Environment*, 1986.
36. Ibid.
37. Ibid., fig. 3.1.
38. D. Vogel, *National Styles of Regulation: Environmental Policy in Great Britain and the United States* (London: Cornell University Press, 1986); L. J. Lundqvist, *Environmental Policies in Canada, Sweden, and the United States: A Comparative Overview* (Beverly Hills: Sage, 1974).
39. *The State of the Environment*, 1985.
40. The relevant international literature includes P. B. Downing, "Cross-national Comparisons in Environmental Protection: Introduction to the Issues," *Policy Studies Journal* 11 (1982): 39–43; P. Knoepfel and H. Weidner, *Handbuch der SO2–Luftreinhaltepolitik*, 2 vols. (Berlin: Erich Schmidt Verlag, 1980); P. Knoepfel and H. Weidner, "Formulation and Implementation of Air Quality Control Programmes: Interest Recognition Patterns," *Policy and Politics* 10 (1982): 85–109; N. Haigh et al., *European Community Environmental Policy in Practice* (London: Graham and Trotman, 1986); L. J. Lundqvist, *The Hare and the Tortoise: Clean Air Policies in the U.S. and Sweden* (Ann Arbor: University of Michigan Press, 1980); Vogel, *National Styles of Regulation* 1986; Enyedi et al., *Environmental Policies in East and West*.
41. M. Jänicke "Blauer Himmel über den Industriestädten—eine Optische Täu-

schung," in *Umweltpolitik: Beiträge zur Politologie des Umweltschutzes*, M. Jänicke, ed. (Opladen: Leske Verlag & Budrich GMbH, 1978), pp. 150–65.

42. Baltic Marine Environment Protection Commission, *Baltic Sea Environmental Proceedings No. 5b* (Helsinki, 1981); OECD *Mediterranean Pilot Study of Environmental Degradation and Pollution from Coastal Development* (Paris: OECD 1976); *Proceedings of the International Conference on Environmental Protection of the North Sea*, organized by the Water Research Center (United Kingdom, March 1987).

43. D. Elsom, *Atmospheric Pollution: Causes, Effects, and Control Policies* (Oxford: Basil Blackwell, 1987).

44. J. E. Carroll, ed., *International Environmental Diplomacy* (Cambridge: Cambridge University Press, 1988).

45. H. Siedentopf and J. Ziller, *Making European Policies Work*, 2 vols. (London: Sage, 1988).

46. Haigh, *European Community Environmental Policy in Practice* 1986; N. Haigh, *EEC Environmental Policy and Britain* (London: Longman, 1987).

47. "Biotechnology and Genetic Manipulation: What Risks, What Advantages," *Institute for European Environmental Policy Bulletin* 39 (May 1987).

48. U. E. Simonis, *Preventive Environmental Policy Concept and Data Requirements* (Berlin: International Institute for Environment and Society, Science Center, 1984).

49. K. von Moltke, *The Vorsorgeprinzip in West German Environmental Policy* (London: Institute for European Environmental Policy, 1987).

50. *Implementation of Directive 82/501/EEC (Prevention of Major Accidents) in the First Five Years after Its Adoption*, report for the European Commission by the Institute for European Environmental Policy, June 1987.

51. European Community Directive No. 85/337.

5

LEGAL ASPECTS OF TRANSBOUNDARY POLLUTION IN EUROPE

Winfried Lang

International cooperation is an essential component of effective environmental protection. The most common argument advanced in favor of merging national efforts cites environmental damage that occurs or is likely to occur outside the territory in which it originates. Transnational threats to the environment shared by two or more states require early and preventive action by the polluting as well as the potential victim states. International action is also needed when the economic and technological capacities of a country are insufficient to handle ecological emergencies. Both developing and industrialized countries may need to rely on foreign assistance to cope with major human-made disasters. It is now widely acknowledged that the financial burden on countries combating pollution could be alleviated by merging funds and legislating a division of labor that goes beyond ad hoc measures.

The 1972 Stockholm Conference on the Human Environment may be considered the point of departure for a new branch of international law: international environmental law. In its final declaration, Principle 21, the conference struck a balance between a state's sovereign right to exploit its own natural resources and its duty to avoid damage that may affect areas beyond its jurisdiction.[1]

One of the features of nascent international environmental law is the scarcity of jurisprudence on which to build. With the exception of the Trail Smelter case, the Lac Lanoux case, and some aspects of the Corfu Channel decision of the International Court of Justice,[2] lawmakers have had to seek guidance in rather vague concepts such as good neighborliness.[3]

Lawyers faced with the need to regulate transnational problems usually resort to treaty making, that is, the elaboration of "hard law" rules. But because of the novelty of transboundary pollution, governments have had to look for new approaches. Drawing on experience acquired in other areas, such as the elaboration of international economic law in the North-South context, the first steps toward firmer commitments were quite frequently "soft law" rules. Although these rules, sometimes referred to as "guidelines," "principles," or "codes of conduct," do not have the binding force of traditional law, they reflect a broad consensus.[4] Parallel to this new way of lawmaking, states have continued to conclude formal treaties and conventions whenever their "consent to be bound" was sufficiently strong to enter into strict obligations.

Experience shows that states willingly accept firm commitments only if they can fully evaluate the burden of duties resulting from a new rule. The likelihood of success of new "strict" or "hard" rules increases if their scope is limited, either in subject matter or in the number of participating states. Treaties that deal with a single type of pollution and that are bilateral or regional stand a better chance of being signed and ratified than treaties that have a broader coverage, such as pollution in general.

Many of the first attempts at creating environmental law have taken place in Europe. In the Final Act of the Conference on Security and Cooperation in Europe, signed at Helsinki in 1975, each participating state acknowledged its duty to ensure "that activities carried out on its territory do not cause degradation of the environment in another state or in areas lying beyond the limits of national jurisdiction." In order to further the cooperation among signatories of the Final Act, it was agreed "to increase the effectiveness of national or international measures for the protection of the environment" and "to take necessary measures to bring environmental policies closer together and, where appropriate and possible, to harmonize them." At the same time the participating states confirmed their intention to promote "the progressive development, codification, and implementation of international law as a means of preserving and enhancing the human environment."[5] It must be recognized with satisfaction that the first multilateral legal instrument covering air pollution—the Convention on Long-Range Transboundary Air Pollution, signed in 1979—was an offspring of the Helsinki conference process.

Before turning to some specific types of pollution or dangerous activities covered in Europe by legal instruments, it may be helpful to list briefly the objectives that, according to experts,[6] should determine the content of environmental lawmaking:

1. States should refrain from activities within their jurisdiction and control that damage the environment of other states or areas outside the limits of national jurisdiction.
2. Costs of damage from transnational pollution should be internationalized, that is, they should be borne by the polluting rather than the victim state.
3. In any case of international use of natural resources, a state should take into account a neighboring polluted state's interests and reconcile them with its own interests.
4. In its cost-benefit analyses a state should not discriminate between pollution affecting its territory and population, and pollution affecting the environments of neighboring states.

With these objectives in mind we can now scrutinize some European achievements. To illustrate the application of a variety of legal instruments to which the states of Eastern and Western Europe are a party, this analysis focuses on three principal concerns: air pollution, water pollution, and the transboundary effects of ultrahazardous activities.

AIR POLLUTION

Most of the area of southern Sweden and Norway has acid-sensitive soils and lakes. Acid precipitation occurs when airborne sulfur and nitrogen oxides, emitted primarily by power plants and industrial processes, combine with moisture in the air to form sulfuric and nitric acids. A similar effect is created when dry sulfate particulates combine with moisture on the ground. As a result, aquatic ecosystems, crops and forests, and the surfaces of stone buildings are damaged. Swedish scientists have determined that up to one-third of the sulfur compounds in Sweden's atmosphere originate in Eastern Europe, primarily the German Democratic Republic (GDR), Poland, and Czechoslovakia. Norway has been most affected by sulfur emissions from Great Britain and the Federal Republic of Germany (FRG).

Controlling acid precipitation in Scandinavia will probably require the cooperation of East and West European states. Since the late 1960s, Scandinavia has been seeking an international agreement to prevent further acidification of its lakes and soils.[7] An opportunity to pursue such an agreement arose in 1975 with the Helsinki conference, when the Soviet Union proposed convening all-European congresses devoted to three pressing problems affecting all of Europe: energy, transportation, and the environment. Because the environment was considered to be the most innocuous of the three, the Economic Commission for Europe, a regional body of the United Nations charged with the economic follow-up to the Helsinki conference, embarked upon the preparation of a "high-level meeting on the environment." Some Western states argued that any major East-West conference on the environment should be a forum not only for the delivery of general statements of good will but also for the conclusion of a concrete agreement. The Scandinavian nations, most affected at that time by acid rain, and several East European states, keen to give more substance to detente, shared an interest in signing an international convention on the environment. Other countries in the commission, such as the FRG and Great Britain, resisted these efforts, preferring instead to start with a soft-law instrument, such as a declaration of principles, rather than immediately making firm commitments. Nonetheless, in 1979 a treaty on transboundary air pollution was signed by thirty-three states of the European Economic Community and of Western and Eastern Europe, as well as the United States and Canada. The vague wording of most provisions of this treaty is due in part to a late shift from recommendatory to mandatory language.

In the treaty, the contracting parties declared themselves "determined to protect man and his environment against air pollution"; they "shall endeavor to limit and, as far as possible, gradually reduce and prevent air pollution" (Article 2).[8] Furthermore, they envisaged the development of policies and strategies that would combat the discharge of air pollutants (Article 4). The most innovative provision was incorporated into Article 5: consultations were to take place at an early stage between the contracting parties affected by or exposed to a significant risk of long-range transboundary air pollution and the contracting parties within whose boundaries and subject to whose jurisdiction significant long-range transboundary air pollution originates, or could originate, in connection with activities carried on or contemplated there. This duty of advance consultation is a major contribution to the development of international environmental law. From a legal standpoint it represents the preventive exercise of a state's right to the diplomatic protection of its nationals. Furthermore, each contracting party committed itself to developing the best policies and strategies for managing air quality compatible with balanced development. This commitment suggested that economic development was still considered more important than ecological concerns. The obligation to use the best available technologies to avoid air pollution was restricted by a reference to the economic feasibility of such technologies. The remaining provisions addressed research and development, exchange of information, and the establishment of a monitoring program called the Cooperative Program for Monitoring and Evaluation of Long-Range Transmission of Air Pollutants in Europe."

Apart from Article 5 on consultations, the convention contains no concrete measures contributing to the improvement of air quality in Europe but only proposes a framework and machinery for cooperation, which may gradually facilitate the adoption of concrete measures. Indeed, a first concrete, cooperative measure, agreed upon in 1985, was an additional protocol to the convention in which the contracting parties committed themselves to a 30 percent reduction of their national annual sulfur emissions or their transboundary fluxes not later than 1993, using 1980 levels as the basis for calculation.[9] Although not all European states have become parties to this protocol—the GDR, Poland, Romania, Greece, Italy, Portugal, Spain, Switzerland, and Great Britain have not joined—it represents a significant step on the road toward better air quality in Europe. A further step was taken when a similar agreement regarding emissions of nitrogen oxides was signed in Sofia on 1 November 1988.[10]

WATER POLLUTION

Water pollution was one of the first environmental problems to receive international attention. Numerous bilateral treaties covering boundary waters have already been concluded. For instance, treaties signed by Austria with Yugoslavia, Hungary, and Czechoslovakia make special provision for early warning in the case of pollution accidents affecting boundary waters.[11]

The Helsinki Convention on the Protection of the Marine Environment of the Baltic Sea Area is certainly the most important legal instrument dealing with water pollution from an East-West perspective.[12] According to this treaty implemented in 1980 by Denmark, Finland, the GDR, the FRG, Poland, Sweden, and the USSR, the contracting parties are bound to prevent and reduce pollution and to protect and improve the marine environment of the Baltic Sea area. Included are a number of subsidiary obligations; for example, significant quantities of substances such as mercury, cadmium, lead, and zinc; persisting pesticides; oil; and radioactive materials are not to be introduced into the marine environment without a special permit. Apart from preventing land-based pollution, special provision is made to combat pollution from ships in an annex to the convention devoted to controlling discharges of oil and noxious liquids carried in bulk, and of sewage and garbage from ships. Permissible dumping in the Baltic Sea is restricted to exceptional cases, such as the dumping of dredged spoils that do not contain significant quantities of any of the substances referred to above. At the same time, the contracting parties are called upon to take all appropriate measures to prevent pollution resulting from the exploration and exploitation of the seabed and its subsoil.

The Baltic Marine Environment Protection Commission was established to review the implementation of the treaty, define criteria for pollution control, and recommend concrete measures to further the objectives of the convention. This commission is an international organ of the most traditional type in that its exclusively intergovernmental character requires that all initiatives be approved unanimously by all member states. The rather conservative character of the commission is most likely due to the ideological diversity of the member states of the convention.

ULTRAHAZARDOUS ACTIVITIES

Agreements relating to the international consequences of nuclear accidents have been concluded in Western Europe since the early 1960s. In 1963 the four Scandinavian countries and the International Atomic Energy Agency (IAEA) signed the Nordic Mutual Emergency Assistance Agreement in Connection with Radiation Accidents. Bilateral treaties have been negotiated between France and Belgium, France and Switzerland, and the FRG and some of its West European neighbors (such as Switzerland, France, the Netherlands, and Denmark).[13]

The first East-West agreement on radiation accidents was concluded between Austria and Czechoslovakia in 1984 and to some extent served as a model for a treaty negotiated between Austria and Hungary in 1987.[14] The 1984 agreement provides for a system of information sharing that operates on three levels. First, a general exchange of information covers matters such as the national program of energy production and safety rules for nuclear power stations. On the second level, at least six months before the start-up of a nuclear power station located close to its border, Czechoslovakia must inform Austria about specific safety measures operative at that station. Also, radiation levels monitored in the vicinity of the border are reported regularly to the other side. The third level of information relates to emergencies: any unforeseen event that may endanger the population living immediately across the border is to be reported to the other side. Even though this agreement provides for cooperation in ecological crisis management, it does not fully meet the expectations of many Austrians who had hoped to prevent the construction of new nuclear power stations close to Austria's borders. In 1978 the people of Austria, by voting in a popular referendum, vetoed the opening of the first nuclear power plant in their country. This decision was confirmed by the Austrian parliament when it renounced all nuclear activity aimed at the production of energy.

A multilateral system of crisis management was not in place before the nuclear accident at the Chernobyl plant in April 1986. Guidelines on reporting events and integrating planning and information exchange in the case of a transboundary release of radioactive materials had been drafted in 1982–84 by a group of experts of the IAEA. This set of soft-law rules was complemented by guidelines for mutual emergency assistance.[15] None of these guidelines, however, acquired the force of law, because the necessary political will was lacking. Still, by their mere existence these guidelines facilitated the work of two groups of experts convened by the IAEA in July 1986 to draft two universally applicable conventions, one devoted to early notification and the other to mutual assistance. The long-range transboundary effects of the Chernobyl accident generated the political will for the adoption of those conventions. Widespread popular concern over the consequences of the accident proved enough to override considerations of national sovereignty as well as the long-standing opposition of the nuclear industry to such agreements.

The Convention on Early Notification, signed on 26 September 1986, has entered into force for twenty-five states including all countries of Eastern Europe except Romania.[16] It contains two main obligations: first, any nuclear accident must be reported directly or through the IAEA to states that may be physically affected by it; second, following early notification, the state in which the accident occurred is to provide detailed information on the causes of the accident, its characteristics, meteorological and hydrological conditions, and off-site protective measures taken or planned. These obligations apply to all nuclear accidents except those that are caused by nuclear weapons. In the case of weapons-related accidents, the obligation to notify was abandoned in favor of the simple possibility of notification ("may notify"). When signing the convention, all five nuclear powers committed themselves by unilateral declaration to use this option, making compliance with this soft-law element of the convention also likely. Although the convention on early notification has been hailed as a major achievement in the field of environmental protection, it contains an important drawback: international notification is activated neither automatically nor by reference to any objective criteria. On the contrary, it is up to the authorities of the state in which the accident occurs to evaluate whether radioactive material is likely to be released, whether it is likely to be released into another state, and whether that release could be significant to another state. Only if the response to all three questions is in the affirmative is the state obligated to notify potential victim states of the imminent danger. The state experiencing the accident determines the extent of its obligations to notify and inform on the basis of its own evaluation of the consequences.

The Convention on Assistance in the Case of a Nuclear Accident or Radiological Emergency was also signed on 26 September 1986, by twenty states, again including all the East European countries except Romania. This agreement may be considered the second pillar of the system of crisis management erected by the IAEA.[17] Although the initial draft submitted by the secretariat contained an explicit obligation to render assistance ("shall use its best endeavors to render promptly and within the limits of its capabilities the assistance requested"), the final version stipulates only that "States Parties shall cooperate to facilitate prompt assistance." Also missing from the final text are common or at least fully coordinated contingency plans. The treaty does, however, regulate in a relatively detailed manner the modalities of assistance. These include the direction and control of assistance, the reimbursement of costs, the privileges and immunities to be granted to the personnel of the assisting party, and the highly sensitive question of claims and compensation if death or injury, damage to or loss of property, or damage to the environment occur in the course of providing assistance. The initial obligation of the state to which a request for assistance is addressed is minor: this state shall promptly decide and notify the requesting state whether it is in a position to render assistance. Because most participants in the 1986 conference considered this universal convention merely a framework for further action, one may assume that they are willing to accept more far-reaching obligations of assistance toward a smaller number of states. It may be possible to strike a balance between national self-interest and international solidarity by facilitating mutual assistance among states that lie in a specific geographic region or are politically like-minded. One might eventually expect the Helsinki conference to serve as the basis for more ambitious agreements stipulating a stringent duty of states to help one another in cases of environmental emergency.

CONCLUSIONS

The success of international environmental cooperation depends on the economic resources of the countries involved, as well as on the nature of their sociopolitical systems. In this respect, most countries of Eastern Europe find themselves in a considerably different situation from West European states.[18] Numerous East European nations still give priority to heavy industry, with its high energy requirements, and rely to a high degree on brown coal for their energy production— on account of which reliance these countries generate more air pollution than countries that use less polluting energy resources such as oil or natural gas. Many East European countries also rely heavily on chemical fertilizers and pesticides to expand agricultural produc tion—chemicals that are major sources of water pollution. Finally, the environmental policies of Eastern Europe are frequently destabilized by repeated efforts at economic reform while at the same time suffering from the limited flexibility that characterizes most centrally planned economies.

The extent to which international obligations are complied with and environmental concerns are integrated into governmental decision making depends on, among other factors, the sociopolitical system of a country, the respect of its elites for the rule of law,[19] and the role played by public opinion in policy-making. When public discussion of controversial issues is restricted—as it is in many East European countries—environmental problems are more likely to be neglected because environmental groups have fewer opportunities to organize to influence policy. Environmental critics may be suspected of challenging the political system as a whole. Furthermore, in centrally planned economies it is quite common for institutions charged with a particular economic activity (such as mining or production of electricity) to be made the custodians of the natural resources related to that activity. And those who are committed to production targets are likely to disregard environmental concerns, especially if measures to protect the environment seem to jeopardize the fulfillment of the production goals. An independent judiciary, largely absent in Eastern Europe, has helped in some Western nations to strike a balance between economic and ecological interests. Frequently lacking independent environmental groups and an independent judicial system, closed societies of the Communist type are forced to rely almost exclusively on the policy impulse coming from the government and the party.

At the international level the implementation of rules to protect the environment suffers from a number of structural deficiencies. Public opinion at this level is frequently fragmented by political boundaries. Official willingness to cooperate may also be stymied by ideological considerations or perceptions linked to historical experiences of confrontation. Because of the decentralized nature of the international community, the fulfillment of obligations depends largely on the goodwill of governments: their permissive consensus determines the content of new rules and obligations while their assessments of expediency, or of economic feasibility, affect the concrete application of internationally agreed rules. The verification of compliance remains a volatile issue as long as no formal international authority is empowered to penetrate the protective shield of national sovereignty in disregard of the sanctity and exclusiveness of internal matters.

Is there any chance for a truly international system of environmental protection against these tremendous odds? The countries of Europe, despite their economic and political heterogeneity, face identical problems: the problems of industrialized societies. Because of contiguity, they also share a common geography and environmental interdependence. This dual challenge motivates the search for common solutions, or at least a coordinated response. International rules elaborated in a European context are likely to contain more substance and more stringent obligations than legal instruments of a universal character, which have to take into account the somewhat different problems of developing countries.

The improvement of the quality of the environment in Europe requires responsible and effective national policies and the conclusion of ever more international preventive and crisis-management agreements. If these agreements go beyond general declarations of intent and contain stringent and verifiable obligations that are actually applied, it is possible in the long run that they will be merged into a single network of rules and institutions—a "European Environmental Protection System."

NOTES

1. Stockholm Declaration on the Human Environment, *International Legal Materials*, vol. 11, 1972, pp. 1416–21; Louis Sohn, "The Stockholm Declaration on the Human Environment," *Harvard International Law Journal* (1973): 423–515.
2. Trail Smelter arbitration: A Canadian smelting plant was emitting sulfur dioxide clouds, which caused extensive damage in the state of Washington. The tribunal stated that no state has the right to use or permit the use of its territory in such a

manner as to cause injury by fumes in or to the territory of another or the properties or persons therein, when the case is of serious consequence and the injury is established by clear and convincing evidence. *Encyclopedia of Public International Law*, 2 (1981), p. 276. Lac Lanoux arbitration: A dispute between France and Spain related to the diversion of waters from a lake in France, which flow into Spain. The tribunal rejected the Spanish contention that the French scheme required the consent of the downstream state (Spain). The latter's interests were to be taken into account. *Encyclopedia of Public International Law*, 2 (1981), p. 167. Corfu Channel case: A dispute between Britain and Albania on the right of passage through the channel. Two British destroyers forced their passage, struck mines, and were heavily damaged. The International Court of Justice ruled that Albania should have known about the mine laying and should have warned the British warships because every state is under the obligation not to knowingly allow its territory to be used for acts contrary to the rights of other states. *Encyclopedia of Public International Law*, 2 (1981), p. 63.

3. For a general overview see Winfried Lang, "Environmental Protection, the Challenge for International Law," *Journal of World Trade Law* (September–October 1986): 489–96; a detailed account is given by Allen Springer, *The International Law of Pollution* (Westport, Conn., London: Quorum Books, 1983).

4. Winfried Lang, "Die Verrechtlichung des internationalen Umweltschutzes, vom soft law zum hard law," *Archiv des Völkerrechts* (1984): 283–305.

5. Conference on Security and Co-operation in Europe: Final Act, *International Legal Materials*, vol. 14 (1975), pp. 1307–9.

6. Günther Handl, "Managing Nuclear Wastes: The International Connection," *National Resources Journal* 21 (April 1981): 297.

7. Armin Rosencranz, "The ECE Convention of 1979 on Long Range Transboundary Air Pollution," *American Journal of International Law* 75, no. 4 (October 1981): 975–82; Amasa Bishop, "High-Level Meeting within the Framework of ECE on the Protection of the Environment," *Environmental Conservation* (summer 1980): 165.

8. Convention on Long-Range Transboundary Air Pollution, *International Legal Materials*, vol. 18 (1979): 1442.

9. Hendrick Vygen, "Air Pollution Control Success of East-West Co-operation," *Environmental Policy and Law* 15 (1985): 6–8.

10. See ECE document EB. Air/CRP.5 (31 October 1988).

11. *Austrian Federal Gazette* (Bundesgesetzblatt), no. 199/1956, no. 255/1959, and no. 106/1970.

12. Convention on the Protection of the Marine Environment of the Baltic Sea Area, *International Legal Materials*, vol. 13 (1974), p. 546; Boleslaw Boczek, "International Protection of the Baltic Sea Environment against Pollution: A Study in Marine Regionalism," *American Journal of International Law* 72, no. 4 (October 1978): 782–814; Finnish Baltic Sea Committee (Ministry of the Interior), "Man and the Baltic Sea," 1977.

13. These legal texts are reviewed by Thomas Bruha, "Internationale Regelungen zum Schutz vor technisch-industriellen Umweltnotfällen," *Zeitschrift für ausländisches öffentliches Recht und Völkerrecht* (with English summary) 44/1 (1984), pp. 1–65. See also N. Pelzer, "Legal Problems of International Danger Protection and of International Emergency Assistance in the Event of Radiation Accidents," in *Peaceful Uses of Atomic Energy*, UN-IAEA Proceedings Series, vol. 3 (1972), pp. 451–63.

14. *Austrian Federal Gazette* (Bundesgesetzblatt), no. 208/1984.

15. See IAEA-documents INFCIRC/321, *Guidelines on Reportable Events, Integrated Planning and Information Exchange in a Transboundary Release of Radioactive Material*; and INFCIRC/310, *Guidelines for Mutual Emergency Assistance Arrangements in Connection with a Nuclear Accident or Radiological Emergency*.

16. Convention on Early Notification of a Nuclear Accident, *International Legal Materials*, vol. 25 (1986), pp. 1370–76.

17. Convention on Assistance in the Case of a Nuclear Accident or Radiological Emergency, *International Legal Materials*, vol. 25 (1986), pp. 1377–86.
18. John Kramer, "The Environmental Crisis in Eastern Europe: The Price for Progress," *Slavic Review* (summer 1983): 204–20; Christine Zvosec, "Environmental Deterioration in Eastern Europe," *Survey* (winter 1984): 117–41.
19. O. Kolbasov, "Study of the Development of Environmental Law in the Socialist Countries of Eastern Europe," UNEP/IG.28/Background Document No. 8.

6

INTER-GERMAN ISSUES OF ENVIRONMENTAL POLICY

Helmut Schreiber

The tenth of June 1987 was an important day for inter-German environmental policy. On this day the long-awaited framework agreement on environmental cooperation between the Federal Republic of Germany (FRG) and the German Democratic Republic (GDR) was concluded. This was the first such agreement between the FRG and a socialist neighbor. Furthermore, it may set a precedent for future agreements between Western countries and other members of the Council for Mutual Economic Assistance, where the GDR plays an important role.[1] A good relationship between the two German states is an important condition for effective cooperation between East and West.

TRANSBOUNDARY POLLUTION AND THE TWO GERMANIES

The FRG is situated in the heart of Europe and therefore is more affected by transboundary environmental problems than are other European countries.[2] Although transboundary pollution in the area arose in the mid-nineteenth century with the process of industrialization, it became an important item on the European political agenda only recently. For example, as recently as ten years ago, West German officials were not willing to accept any responsibility for the acidification of lakes in the Scandinavian countries.[3] At conferences on transboundary air pollution, West Germany always pointed out that there was no final proof that West German emissions were responsible for the ecological deterioration of the Scandinavian lakes and forests, a position still held today by Great Britain.[4] In the early 1980s, the phenomenon of *Waldsterben* (death of the forest) spread to West Germany itself. This development led to a basic change of attitude in the West German environmental protection agencies. Since then, transboundary air pollution has become one of the major concerns of environmental policy in West Germany and is an accepted topic of international discussion in Europe.

Despite considerable efforts to reduce air pollution during the last ten years, the overall quantity of emissions in Europe is still high. The GDR is a leading emitter of sulfur dioxide, a product of the widespread combustion of its indigenous brown-coal resources; because of the GDR's small territory, an "export" of emissions from the GDR to its neighbors is unavoidable. The GDR's eastern neighbors are affected the most by these exports because of the prevailing westerly winds in central Europe. But often the FRG is also affected, especially during the winter. Over the last few years, periods of severe smog have resulted in West Berlin and regions near the border.

Transboundary environmental problems are not restricted to air pollution but also include water pollution. Several rivers, such as the Rhine and the Elbe, have their sources in other European countries and then cross West Germany on their way to the sea. For instance, the Elbe flows through Czechoslovakia and the GDR before it reaches the FRG. Although the GDR has an impressive body of water law, the Elbe is highly polluted. Ammonium compounds, heavy metals, and chloride hydrocarbons are the worst toxic agents in the river. Ninety-five percent of this river's pollution is imported from these eastern neighbors. Cooperation between the FRG and Czechoslovakia to reduce pollution of the Elbe has begun, but it will bring only minimal results if the GDR is not involved, because 55 to 60 percent of the pollution measured in Hamburg originates in the GDR.[5]

Other rivers have their sources in West Germany or flow through other countries after crossing West German territory—for example, the Danube and the Rhine, in which cases West Germany is often accused of being an emitter. Although the European Economic Community (EEC) has served as a framework for resolving conflicts in the field of transboundary pollution with the FRG's western and northern neighbors, until recently there was no political or legal basis for bilateral or international agreements to handle these problems with the countries of the East bloc.[6]

MAJOR PROBLEMS IN INTER-GERMAN COOPERATION

Historically two major problems have hindered FRG-GDR cooperation in the environmental area. The first problem has been the status of West Berlin. Based on its understanding of the Four Powers Treaty on Berlin of 1971 (*Viermächteabkommen*), the GDR supports the position that West Berlin is an "independent political entity." The GDR, therefore, has not been willing to include the city in its negotiations or agreements with the FRG. The FRG, on the other hand, wanting to strengthen ties between itself and West Berlin, attempts to include West Berlin in all its treaties with the GDR. The issue of Berlin has been of particular importance in the field of environmental protection because since July 1974 the West German Environmental Protection Agency (*Umweltbundesamt*) has been located in West Berlin. The member states of the Warsaw Treaty Organization viewed this move as a violation of the Four Powers treaty, because it implies that West Berlin is a constituent part of the FRG. Consequently the GDR withdrew from all talks on environmental questions with the FRG.[7] This major obstacle to environmental cooperation was overcome only recently.[8] In the future, experts from the *Umweltbundesamt* will be able to participate in cooperative efforts between the two German states; they will not, however, represent the *Umweltbundesamt* but will instead participate as experts in their own right.

The second major problem involves the different approaches taken by the two countries in addressing environmental pollution control. While the FRG, like most countries, follows the "polluter pays" principle, the GDR maintains that the country that benefits from the measures to control pollution should carry the financial burden involved. The GDR's position on this issue may simply reflect economic necessity rather than deep conviction, given the country's hard-currency debt and shortage of investment resources. In practice, the West German government has compromised its polluter-pays principle in its relations with the GDR by subsidizing environmental protection in the GDR. The FRG's willingness to do so reflects the "special relationship" between the two countries and the West German desire to protect the interests of the residents of West Berlin and regions bordering the GDR.

HISTORY OF INTER-GERMAN ENVIRONMENTAL COOPERATION

In 1972 the Treaty on Basic Relations (*Grundlagenvertrag*), one of the most important treaties between the FRG and the GDR, was concluded. In this treaty a basic foundation was laid for relations between the two German states in fields of vital interest, including environmental protection. Supplementary Protocol II (*Zusatzprotokoll II*) specified that "in the field of environmental protection, agreements should be concluded between the Federal Republic of Germany and the German Democratic Republic to prevent damage and dangers for both sides."[9] Talks on these issues began in November 1973 but were cut off by the GDR following the establishment of the FRG's *Umweltbundesamt* in West Berlin. It was not until April 1980 that talks on these problems were resumed.[10] Until the conclusion of the Basic Agreement on Environmental Cooperation in June 1987, these talks mainly addressed specific problems of water pollution: the GDR has agreed to intensify its efforts to control pollution in exchange for financial assistance from the FRG. The talks since 1980 have concentrated on the problem of transboundary water pollution. In September 1982 an agreement was concluded between the two countries that committed the GDR to improving three of its large water-purification plants to help clean up rivers that cross into West Berlin.[11] The West German government, in turn, agreed to subsidize these efforts, paying 68 million deutsche marks to upgrade the purification facilities to help stabilize the ecological balance of West Berlin's rivers and lakes,[12] which are heavily polluted by domestic and chemical sewage from the GDR. A second agreement relates to the installation of a purification plant near the Bavarian border. Here, the pollution of the small river Röden in the GDR had provoked popular protests in West Germany. In 1983, West Germany agreed to pay about 30 percent of the construction costs of new purification facilities.

An unresolved environmental problem between the two German states is the reduction of the salt content of the Werra River, which flows through Hesse after being polluted by salt emissions from potash mining and potash industries in the GDR. In previous years, several agreements have been concluded concerning this river and related problems of the potash industry. Again, West Germany will pay vast sums to clean the pollution caused by the GDR.

The GDR is one of the main emitters of sulfur dioxide in Central Europe.[13] Indigenous lignite, an energy resource of high sulfur content and low calorific value, is the basis for the GDR's energy production. In 1982, sulfur dioxide emissions in the GDR reached 4.9 million tons.[14] Despite the magnitude of this problem, transboundary air pollution has become a topic of bilateral talks between the FRG and the GDR only in the last three years. Unlike the field of water pollution, no agreements on air pollution have been reached so far. But some West German officials are optimistic that in the next few years some arrangements will be made, foreseeing that as with water pollution, West Germany will probably help the GDR to build filters, particularly for desulfurization and removal of dust.

The Basic Agreement on Environmental Cooperation

The Basic Agreement on Environmental Cooperation between the two German states signed in June 1987 represents a major leap forward in the field of environmental cooperation.[15] The agreement sets broad goals and can serve only as a framework for detailed work in specific areas. Details for future cooperative efforts are to be laid out in three-year plans, the first of which has not yet been published. The two parties have designated five priority fields for future cooperation:[16]

1. development of technologies and instruments for reducing and measuring air pollution;
2. examination of causes of forest damage and development of methods for its reduction;
3. development of methods to reduce, utilize, and dispose of wastes;
4. improvement of practical methods in the field of nature protection; and
5. improvement of technologies and practical methods in the field of water protection.

Other Agreements

In addition to the Basic Agreement on Environmental Cooperation, other relevant agreements reached in the same period include the following:

1. An agreement on the operation, control, and maintenance of a facility in the GDR to provide drinking water for the town of Duderstadt in the FRG (3 February 1976).[17]
2. An agreement on cooperation in science and technology (8 September 1987). This agreement, which includes cooperation in the natural sciences, engineering, sociology, and the arts, provides for exchanging information; organizing scientific meetings, conferences, and exhibitions; exchanging scientists, teachers, and researchers; organizing research projects; and sharing research materials, instruments, and equipment.[18]

3. An agreement on the development of a protection screen (8 September 1987). This agreement, requiring the exchange of information in cases of accident, was an important milestone in the development of bilateral relations. As a result of the Chernobyl disaster, both sides agreed to mutual and early notification about nuclear accidents[19] and bilateral consultations on the development of peaceful uses of nuclear power.[20]

4. Agreement on cultural cooperation (8 September 1987), which includes visual and interpretative arts, film, music, German philology, museums, protection of monuments, and other areas. Cultural agreements of this kind often lead to further agreements because they promote understanding.[21]

INTERNATIONAL ISSUES OF COOPERATION ON ENVIRONMENTAL POLICY

The improvement in environmental cooperation between the two German states would not have been possible without the work of international organizations. The United Nations Economic Commission for Europe has been particularly important. The commission maintains several programs concerning environmental protection.[22] One of the most significant is the Cooperative Program for Monitoring and Evaluation of Long-Range Transmission of Air Pollutants in Europe, an outgrowth of the historic Convention on Long-Range Transboundary Air Pollution signed in 1979 by the industrialized nations of Europe and North America.[23] This convention was the first treaty involving countries from both Eastern and Western Europe.[24] During the 1970s it became increasingly evident that transboundary pollution is not only a bilateral but also an international problem, and at several international conferences in Europe environmental issues were prominent. One of these was the Conference on Security and Cooperation in Europe, which took place in 1975.[25] Another key environmental conference took place in Munich in 1984, at which top-level officials responsible for the environment, and air pollution control in particular, met to discuss means of controlling transboundary air pollution. One result of this conference was the enlargement of the so-called 30-percent club, whose members commit themselves to reducing their sulfur dioxide emissions by 30 percent in relation to 1980 levels by the year 1993. On that occasion, the Soviet Union and the GDR joined the club. Transboundary pollution is also an important item on the agendas of several other international organizations whose members include Western countries, including the Organization for Economic Cooperation and Development and the European Community.[26]

Inter-German cooperation on environmental protection cannot be separated from cooperation between the FRG and other socialist countries, especially Czechoslovakia. During the last few years, there have been numerous consultations on environmental problems between the FRG and Czechoslovakia. One of the first regions where *Waldsterben* became visible was the Bavarian Forest (*Bayerischer Wald*) along the FRG's border with Czechoslovakia, not far from the sites of huge coal-fired power plants on the Czechoslovak side. These power plants emit large columns of gaseous pollutants, and West German officials have determined that the "death" of the Bavarian Forest is at least partly due to these emissions from Czechoslovakia.

Since 1982 transboundary air pollution has become a topic of several meetings between experts from the two countries. One result has been a basic agreement on environmental protection between West Germany and Czechoslovakia, which should be signed in the near future. So far, there have been no agreements on specific problems between the two countries such as those that exist between the FRG and the GDR.

PROSPECTS

What are the prospects for inter-German environmental cooperation? Successful cooperation depends on the application of measures for economic and technological modernization aimed at realizing environmental goals. In other words, environmentally "friendly" technology must replace, or at least upgrade, the quality of existing equipment. This need is especially important in regard to energy production, where older plants contribute disproportionately to emissions of sulfur dioxide. In the coming years, some 20 billion deutsche marks will be spent in the FRG on controlling air pollution. To match this, the GDR (whose air pollution is worse than the FRG's) would have to spend some 10 billion deutsche marks,[27] although it will not be able to spend even a fraction of this amount. An improvement in environmental quality in the GDR and other socialist countries will depend mainly upon the modernization of the economy.

Both the GDR and the FRG have shown an interest in cooperation in the field of nuclear safety. In the GDR, concern over nuclear safety in the wake of the catastrophe at Chernobyl has led some citizens to question their country's ambitious plans to increase the use of nuclear energy. Cooperation with the FRG in addressing safety issues is one option available to the GDR. On the West German side, the nuclear power industry sees cooperation with the GDR and other countries of Eastern Europe as one way to ease the effects of the slowdown in nuclear construction plans in the West. Meanwhile the West German government hopes that cooperation will help to raise safety standards in the East. During the last few months, several plans for cooperation between the FRG and the GDR have been discussed publicly.

In the last fifteen years, awareness of the limits of the ecological system has grown considerably in Europe and especially in the two German states. It has become increasingly evident that environmental problems can be solved only internationally. The FRG, situated in the center of Europe, exports and imports environmental pollution to and from its neighbors. In the 1970s it discussed transboundary environmental problems mainly with its western and northern neighbors. In the 1980s more attention has been paid to its eastern neighbors. In the last six years considerable progress has been made in combating environmental problems between the FRG and the GDR, particularly water pollution. In other fields such as air pollution, waste management, and the safety of nuclear power plants, discussions have just begun. The Basic Agreement on Environmental Cooperation between the two German states may be an important milestone for improving environmental quality in the heart of Europe. Basic improvements in the environmental situation will be possible only if environmental and economic concerns are linked.

NOTES

1. See "Psychologisch wertvoll," *Handelsblatt*, 11 June 1987.
2. For more details see Gregory Wetstone and Armin Rosencranz, *Acid Rain in Europe and North America* (Washington, D.C.: Environmental Law Institute, 1983), p. 45.
3. For details see Swedish Ministry of Agriculture, *Acidification Today and Tomorrow*, prepared for the 1982 Stockholm Conference on Acidification of the Environment.
4. Wetstone and Rosencranz, p. 66.
5. Jiri Slama, *Umweltprobleme in Osteuropa im internationalen Vergleich* (Munich, 1987), pp. 55–56.
6. Despite the framework for conflicts, there are several unsolved problems like the pollution of the Rhine by France. See Hans Ulrich Jessurum d'Oliveira, "Das Rheinchloridabkommen und die EWG," *Recht der Internationalen Wirtschaft* (1983), p. 322.
7. Wilhelm Bruns, 1984. "Deutsch-deutscher Umweltschutz. Notwendig und möglich," *DDR-Report*, no. 2 (1984); Michael von Berg, "Zum Umweltschutz in Deutschland," *Deutschland-Archiv*, no. 4 (1984).
8. After 1980, the GDR signed basic agreements in the field of environmental protection with the Scandinavian countries, Czechoslovakia, and Poland. See Gesamtdeutsches Institut, 1985, *Zusammenarbeit der DDR mit den Skandinavischen Ländern im Umweltschutz* (Bonn: Gesamtdeutsches Institut, 1985); Bert Dörner and Thomas Leinkauf, "Technologien zum Schutz der Wälder und für reinere Luft," *Berliner Zeitung*, 12 December 1984, p. 9; Helmut Schreiber, "Umweltprobleme

DDR-Polen," Gesellschaft für Deutschlandforschung, ed., *Umweltschutz in beiden Teilen Deutschlands* (Berlin: Humbolt, 1986).

9. See the "Grundlagenvertrag" between the two German states (21 December 1987), Appendix II ("Zusatzprotokoll II"), no. 9, Article 7.

10. *Bulletin des Presse- und Informationsamtes der Bundesregierung*, No. 46/1980.

11. For more detailed information on water problems in the GDR, see Gesamtdeutsches Institut, *Wasserhaushalt und Gewässerschutz in der DDR* (Bonn, 1983); and *Wasserhaushalt, Gewässerschutz in der DDR* (Bonn, 1986).

12. *Bulletin des Presse- und Informationsamtes der Bundesregierung*, No. 89/1982.

13. For details of air pollution problems in the GDR see Hannsjörg F. Buck and Bernd Spindler, "Luftbelastung in der DDR durch Schadstoffemissionen. Ursachen und Folgen," *Deutschland-Archiv* 9 (1982).

14. See Deutscher Bundestag, ed., *Materialien zum Bericht zur Lage der Nation im geteilten Deutschland 1987*, Deutscher Bundestag, 11 Wahlperiode, 11/11, p. 301.

15. The full title is "Vereinbarung zwischen der Regierung der Bundesrepublik Deutschland und der Regierung der Deutschen Demokratischen Republik über die weitere Gestaltung der Beziehungen auf dem Gebiet des Umweltschutzes."

16. Basic Agreement on Environmental Cooperation, Article 2.

17. "Vereinbarung zwischen der Regierung der Bundesrepublik Deutschland und der Regierung der Deutschen Demokratischen Republik über den Betrieb, die Kontrolle in die Instandhaltung der auf dem Territorium der Deutschen Demokratischen Republik gelegenen Teil der Trinkwasserversorgungsanlagen der Stadt Duderstadt."

18. "Abkommen zwischen der Regierung der Bundesrepublik Deutschland und der Deutschen Demokratischen Republik über die Zusammenarbeit auf den Gebieten der Wissenschaft und Technik," Article 5.

19. "Abkommen zwischen der Regierung der Bundesrepublik Deutschland und der Deutschen Demokratischen Republik über Informations- und Erfahrungsaustausch auf dem Gebiet des Strahlenschutzes," Article 1.

20. Ibid., Article 3.

21. See "Abkommen der Regierung der Bundesrepublik Deutschland und der Regierung der Deutschen Demokratischen Republik über kulturelle Zusammenarbeit."

22. Wetstone and Rosencranz, p. 140.

23. This convention was the result of the Helsinki Conference on Security and Cooperation in Europe.

24. The 1979 Convention on Transboundary Air Pollution. UN/EC/GE 79-42960.

25. For more details see Wetstone and Rosencranz, p. 140.

26. Volker Prittwitz, 1984. *Umweltaussenpolitik* (Frankfurt: Campus, 1984), p. 117.

27. M. Schmitz, "Der ungeteilte Dreck: Saubere Luft braucht die Kooperation von Bundesrepublik und DDR," *Die Zeit*, 3 June 1987.

7

ENVIRONMENTAL ISSUES OF INTERNATIONAL WATER MANAGEMENT: THE CASE OF THE DANUBE

Imrich Daubner

Maintaining the purity of water, or at least halting its deterioration, is among the most important contemporary environmental problems. Its resolution depends on comprehensive international cooperation. The decisions adopted by the United Nations Economic Commission on Europe and its Water Committee, and by the Council of Mutual Economic Assistance are of great significance to Europe as a whole. The Final Act of the Conference on Security and Cooperation in Europe (Helsinki 1975) also assigns high priority to problems of environmental protection, including water pollution.

Many countries share the benefits of a common natural resource and must also share the consequences of its deterioration. This circumstance is particularly true of international rivers, which constitute the largest supply of usable water. Developing surface-water resources is expected to be the principal method of meeting the growing demand for water in the future. Increased exploitation of rivers changes their quality and in consequence the distribution of water among users. The ensuing pollution is of vital importance not only because rivers and the adjacent groundwater basins are sources of drinking water, but also because hazardous pollutants accumulated in agricultural crops may seriously affect the quality of food products and therefore the health of the population. Lasting and complex chemical pollution of river water may also lead to marked and often irreversible changes in natural ecosystems.

In this context, the Danube, the second-largest watercourse in Europe, is extremely important internationally. The Danube passes through eight countries: the Federal Republic of Germany (FRG), Austria, Czechoslovakia, Hungary, Yugoslavia, Bulgaria, Romania, and the Soviet Union. Its tributaries flow into four additional countries: Switzerland, Italy, Poland, and Albania. These twelve countries have a combined population of 484 million, and approximately 80 million people live in the drainage area of the Danube. Nine cities with populations of more than 100,000 are situated along the Danube.

Of all the great European rivers, the Danube is the only one that flows from west to east, and for this reason, too, it is economically very important. Its role as a major European waterway will increase even more in the next few years, once the Rhine-Main-Danube shipping canal is completed and vessels can travel from the Black Sea to the North Sea. The Rhine and the Danube will be the two vital arteries linking the industrial areas of sixteen highly developed countries with a combined population of about 560 million.

The Danube is approximately 2,888 kilometers long, draining waters from a catchment area of 817,000 square kilometers in the southern part of Central Europe, discharging them into the Black Sea. Its longest headstream, the brook Brege (47.6 kilometers long), comes together with the second headstream, the Brigach (42.7 kilometers long), at Donaueschingen, where the body of water becomes the Danube. The Danube is divided by mountain ranges into three stretches of approximately 900 kilometers each: the Upper Danube from the headwaters to the mouth of the Morava River near the Austrian-Czechoslovak border, the Middle Danube from the Morava's mouth to Iron Gate Gorge, and the Lower Danube from there to the Black Sea. Its width ranges at low water from 50 meters to 1,200 meters and at maximum high water from 83 meters to 2,800 meters, its depth from 0.8 meters to 2.7 meters. Its surface velocity is between 0.6 and 2.65 meters per second. High-water levels in the spring and summer follow the inflow of water from melting snow and precipitation. The total discharge of the Danube into the Black Sea amounts to about 6,500 cubic meters per second. Because of the inflow of cold Alpine water, in July the average temperature in the Upper Danube fluctuates between 16 and 17 degrees Celsius, in the Middle Danube between 18 and 22 degrees, and in the Lower Danube between 22 and 23.5 degrees. The period of ice cover averages seven days in the Upper Danube, twenty-eight days in the delta region, and eighty-four days in the Middle Danube.

Because of the construction of hydroelectric power plants, the bed load of the Upper Danube has greatly decreased. On the other hand, the suspended load at the mouth of the river is on average 67.5 million tons per year, or 2,140 kilograms per second. The high degree of turbidity in the summer months, which is caused mainly by mineral matter in suspension, causes some reduction in photosynthesis. Special conditions prevail in the delta of the Danube, the second-largest delta in Europe, which covers 5,640 square kilometers. About 80 percent of the delta (4,400 square kilometers) is situated in Romania and 20 percent in the Ukraine (1,240 square kilometers). Approximately half of the delta (51 percent) is below sea level. Only 84 percent (4,900 square kilometers) of its total area is used.

The Danube has always played a significant role in the life and history of the peoples living along its banks. A border in Roman times, it was of vital importance both economically and politically. The Romans set up a series of encampments on its right bank, protected by bridgeheads on the left bank. The most important cities on the Danube of Roman origin are Regensburg, Passau, Linz, Vienna, Bratislava, Komárom, Budapest, Belgrade, Turnu-Severin, Lom, and Reni.

This historical background suggests the complex problems of research and management connected with the Danube. By the same token, the solution to these problems is a common concern of all the riparian countries. Among the most important are the monitoring of water quality, self-purification processes in the river, determination of maximum permissible concentrations of toxic and other chemical substances, accumulation of radioactive and other injurious substances, water supply, wastewater reception, irrigation, fisheries, hydroelectric power generation, flood protection, navigation, and recreation. The multiple use of this vast natural resource has developed gradually over the past millennium, in each country along the entire length of the Danube.

The multipurpose use of the Danube has called for careful coordination between countries. At the Congress of Vienna of 1815, the principle of free navigation was stated. A convention signed in Belgrade on 18 August 1948 established a regime for navigation on the Danube, including port and navigation charges as well as conditions for merchant shipping. Standard principles and procedures for applying these charges and conditions have been developed for all vessels using the river, and the International Commission for the Danube headquartered in Budapest is responsible for enforcing them. Today, ships flying the colors of some twenty-five countries use the Danube annually.

HYDROPOWER AND THE ENVIRONMENT

The Danube contributes significantly to meeting the constantly growing demand for electricity. It falls a total of approximately 700 meters and has a mean discharge of 6,500 cubic meters per second. Its planned maximum capacity of hydroelectric generation at twenty-seven power plants is 46,900 gigawatt hours per year.

The first hydroelectric power plant on the Danube, at Kachlet near Passau in the FRG, began operation in 1927. Since then, many hydroelectric power plants have been constructed, and others are being completed. Construction of power stations has also begun in the countries along the Lower Danube. In 1971 the fifth largest hydroelectric power plant in the world was completed and put into operation at the Iron Gate in the Carpathians (942 kilometers upstream from the mouth) near the Romanian-Yugoslav border. The plant generates 10,700 gigawatt hours annually, and the waters of the Danube are backed up over a distance of 273 kilometers.

At present, a common Czechoslovak-Hungarian power plant system, Gabčikovo-Nagymáros, is under construction. (A similar common project to generate hydroelectric power on the Danube exists between Bulgaria and Romania.) The Gabčikovo-Nagymáros system is one of the largest projects currently under construction in Czechoslovakia. Its scope will without doubt have a strong effect on the environment in the territories along the Danube. Therefore, when the project was still in the planning stage, both its known and hypothetical impacts on the biological conditions of the landscape were being assessed in an attempt to restrict harm to the natural environment. Acting on a resolution by the government of the Slovak Socialist Republic, the chief designer of the power plant—the Hydroconsult Establishment in Bratislava—requested the State Institution for Urban Development and Layout Planning in Bratislava to conduct a study, "The Biological Project for the Territory Affected by the Gabčikovo-Nagymáros Powerplant System." The study was carried out in cooperation with the Institute of Experimental Biology and Ecology of the Center for Biological-Ecological Sciences of the Slovak Academy of Sciences in Bratislava.

The first section of the study provides territorial projections relating to vegetation, distribution of animal groups, hydrobiological-ichthyological conditions, quality of surface water and groundwater, natural and endemic sources of infection, and additional data on public health. The second section provides suggestions for measures aimed at a maximum preservation of natural communities and of the overall character of the landscape. A third part discusses the design of the Biological Project for the affected area and presents targets for its implementation. The draft study is more than four hundred pages long, including maps and other documentation.

The Biological Project itself may be divided into seven main parts:

1. *Changes in documentation.* Changes and additions are proposed to the documentation regarding the impact of the power plant on settlements and recreational and public facilities.
2. *Changes in forest management.* Areas, stages, and principles for removing forest stands are established, and methods provided for maintaining deforested soils. Based on projected changes in groundwater levels and soil types, forests have been categorized according to the level of specific management they will require after the power plant is completed. The document also proposes the afforestation of desiccated Danube branches and areas temporarily requisitioned for construction.
3. *Organization of agricultural production under the groundwater regime.* Areas have been marked off in which conditions for agricultural production will be neither affected, nor improved, nor worsened. A system of protective dams and seepage and drainage channels has been proposed for areas expected to be permanently waterlogged. Also included are proposals for recultivating temporarily requisitioned areas and areas devastated in the process of construction, designating areas for permanent cultures (orchards, vineyards) and grass stands, and using the outer sides of the new dams for agricultural production. To pursue long-term research on the effects of the Danubian Water Scheme on agricultural production, twenty-one experimental areas were designated.
4. *Fishing.* A proposal to establish fish-breeding facilities and to regulate the bed and the shore for an optimum composition of the aquatic ecosystem was elaborated.
5. *Hunting.* Proposals include a territorial layout of hunting districts and their regulation in view of biological-ecological requirements of individual species. The documentation suggests international agreements for the formation and protection of roosting and gathering places for migratory birds.
6. *Sanitary conditions.* Sources of pollution of the Danube and its tributaries on Czechoslovak territory include communal and industrial trial wastewaters, with a high content of pathogenic germs and viruses, and oil pollution. Possible changes in the hydrological regime after the construction of the power plant are taken into ac-

count. In view of the changes in the groundwater regime, plans have been made to create a central water supply for the population in this area, neutralize and recycle wastes, and build a network of water pipes and canalization sufficiently ahead of schedule. This section also deals with protecting workers from sources of infectious disease (such as a flood species of gnats) during construction.

7. *Nature protection.* The program elaborates a new concept of protecting nature in light of expected changes in the natural environment resulting from the construction of the power plant. It includes suggestions for documenting and protecting animal species living on the territory to be permanently flooded, as well as suggestions for the establishment of biotopes in order to recreate an ecological balance after the plant is built.

The creation of the Biological Project is especially significant because it has involved the cooperation of biologists and engineers. It was the first such undertaking in Czechoslovakia and therefore had no benefit of previous experience. Its conclusions and suggestions will certainly require some correction both during and after construction, as technical data on changes in the abiotic component of the landscape become available. The presumption is that cooperation between biologists and technical personnel will continue. Experience elsewhere indicates that the views of biologists, ecologists, and relevant sanitation and public health experts should form an inseparable component in the planning of engineering projects that affect the environment.

OTHER USES OF THE DANUBE'S WATER

The water of the Danube is intensively used for industry and agriculture and, after purification, as drinking water in two countries. Industrial enterprises located on the Danube use especially large amounts of water for cooling and technology. They include steam power plants, refineries, breweries, the iron and steel industry, the chemical industry, dockyards, sugar production, mines, paper factories, and some smaller enterprises.

There are many installations for irrigation along the Danube, but still fewer than what is needed. The total area in need of irrigation is generally estimated at 2.645 million hectares. In the dry regions along the Lower Danube, which suffer from insufficient precipitation in the summer and fall, irrigation alone could triple agricultural production. Forecasts suggest that in future the demand for water for irrigation will rise to twenty-five hundred cubic meters per second from the current thirteen hundred. During the low-water period in the autumn months, the requirements of irrigation will conflict with other uses, particularly shipping. There are no substantial problems with the quality of water for irrigation projects, however, because the rich nutrient content of the water has a beneficial effect. But problems do exist with oil and communal wastewater pollution.

Since 1960 only the city of Budapest, with its more than two million inhabitants, has taken water for municipal and domestic use directly from the Danube. The water plant situated on Margaret Island pumps more than three hundred liters per second into the supply network. Water to supply the needs of the province of Württemberg is taken directly from the river downstream from the Leipheim Dam on the Upper Danube in the Federal Republic of Germany. Other cities— for example, Vienna and Bratislava—take filtered water, which is of very good quality and taste, from the aquifers paralleling the Danube or its tributaries.

The Danube still has one of the largest fish populations of any river in the world, which is increasingly affected by pollution, especially with crude oil products. More than 75 percent of its more than a hundred species of fish are exploited commercially and by anglers. Fishing on the Danube was once a very productive activity, but since controls were introduced a century ago, the size of the catches has declined steadily. Today the average commercial yield is about forty-five hundred tons per year. About three-quarters of it are taken by the countries along the Lower Danube: Yugoslavia, Bulgaria, Romania, and the Soviet Union. A further forty-five thousand tons of fish, mainly cyprinids, are caught in the ponds and lakes of the floodplains and in the delta. In the countries along the Upper Danube almost all fishing is for sport. Here the chief species is beaked carp (*Chondrostoma nasus*), which represents approximately 50 percent of the catch, followed by barbel (*Barbus barbus*) with 10 to 25 percent, pikes, and some salmonids. The fry are placed in the reservoirs of the hydroelectric power plants on the river because the dams have no fish ladders. On the Lower Danube, the sterlet is qualitatively the most valuable fish, whereas quantitatively the cyprinids predominate. The mouth of the Danube is a good environment not only for freshwater fish but also for those moving upstream from the Black Sea. The predominant species here are sterlets.

PROBLEMS OF WATER QUALITY

Many industrial plants as well as large cities along the Danube discharge wastewaters into the river. Pollution is most serious in the

Upper Danube because more factories are situated there and the river carries less water. All the countries concerned are making great efforts to reduce stream pollution either by modifying technological processes or by building specific treatment plants. Vienna has a complex plant for the treatment of water, and similar installations are under construction for Bratislava and other cities. They are urgently needed to treat communal wastewater and waste from various pulp and paper plants, refineries, and chemical plants. It is especially necessary to control the discharges of pathogenic microorganisms and toxic substances.

The treatment of wastewater is especially important because the Danube drainage basin and the river itself belong to one of the most beautiful resort areas in Europe. The variety and individual character of landscapes offer recreation and water sports (hiking, swimming, boating, and hunting). There is hardly a finer experience than to spend time in the river meadows of the Danube, where the last vestiges of Europe's primeval forests survive.

Despite the widely differing uses to which the Danube is put, most sections of the river contain water of grade 2 on a four-grade saprobiological scale. Downstream from most cities pollution is more serious, however, because the majority of wastewaters discharged into the river are inadequately treated, at best. Inadequate treatment gives rise to eutrophication; excessive growth of plankton, bacteria, and fungi; excessive or deficient oxygen content; increased levels of ammonia; and high concentrations of detergents or oil products. These sections of the river are also more heavily contaminated with disease-causing agents such as protozoa, pathogenic bacteria, and enteroviruses, which are permanently present. Consequently, downstream from centers of population or from discharge points for industrial wastes the quality of water drops to grade 3 or even 4. But because of the self-purification capacity of the richly oxygenated and turbulent stream, and also because the wastewaters are usually very diluted in the receiving bodies of the Danube, in most cases the water quality improves quite rapidly within a few kilometers.

INTERNATIONAL COOPERATION

Problems of international water management are extremely critical. As urban and industrial centers continue to expand and grow more dense, the demand for water increases as well. Large amounts of water continue to be drawn off to irrigate agricultural land. The growth of shipping and the consequent development of the Danube as one of

Europe's principal waterways, the construction of power plants, and growing domestic and industrial discharges all contribute to the deterioration of the river. Most of the demands on the Danube are at variance with its function as an indispensable source of recreation for the people in the countries along its course, and for this reason some years ago special attention was devoted to the Danube on an international level. In 1964 the World Health Organization (WHO) and its regional office for Europe launched a study of common actions that needed to be taken. These were consolidated and passed by a "Workshop on the Study and Assessment of the Water Quality of the Danube" convened in Copenhagen in 1975. In December 1977 the WHO's regional office for Europe, together with the United Nations Development Program and the Hungarian National Water Authority, cosponsored a seminar in Budapest on the topic of "Pilot Zones for Water Quality Management." Among the conclusions reached at the seminar were these:

1. The Danube is experiencing the effects of very rapid socioeconomic development. In all Danubian countries, growing industrial and agricultural production and urbanization increase the need for water. Most of this water will have to come from the Danube, and therefore any further deterioration in its quality may be harmful to human health. To minimize this possibility, international cooperation is required.
2. Problems of research and management related to the Danube and its tributaries are fundamental, and decisions being made currently will inevitably affect the river both in the short and long terms.
3. Sediment transport, with associated bioresistant substances such as heavy metals and polyaromatic hydrocarbons, is a major long-term problem with potential effects on both human health and the ecosystem.
4. Conflicts exist between upstream and downstream interests, and they must be taken into account in developing international cooperation.
5. Analytical, hydrological, physical, chemical, radiological, and biological data are stored in many countries. Some of these data would be very useful for the construction of mathematical models. Exchange of these data could be an early step in international cooperation, after a common format and coding system are designed.

A conference of representatives of all the countries on the Danube took place in Bucharest on 11–13 December 1985. Governmental delegations as well as representatives of some international organizations were present. The main purpose of the meeting was to adopt a declaration of cooperation between Danubian countries on water economy and water protection. The declaration addressed problems of exchanging information among the riparian countries, establishing some economic balances, and elaborating a system to estimate water quality.

The countries along the Danube have conducted research for several decades. At first it consisted mostly of hydrological observations, but later it was expanded to physical, chemical, biological, radiological, microbiological, and virological studies. These studies were initially undertaken more or less independently in individual countries, without mutual information and coordination. The thirteenth Congress of the International Association for Limnology (SIL), held in Helsinki in 1956, adopted a resolution to establish an international organization that would link specialists involved in research on the Danube and its tributaries. Subsequently, the International Working Association for Danube Research (*Internationale Arbeitsgemeinschaft Donauforschung* [IAD]) was founded, also in 1956, with Vienna as its seat. Its members include all the countries lying on the Danube and Switzerland, which is connected with the Danube by a tributary, the Inn. IAD is financed by annual contributions from member countries as well as SIL, into which IAD was incorporated in 1959.

The aim of IAD is to coordinate scientific data on the Danube and its tributaries. All institutions and persons who conduct such research or are interested in its results are invited to participate. Because IAD operates on a voluntary basis, it is able to work effectively while at the same time respecting the positions of its participants (which include academies of sciences, universities, ministries and their research institutes, and factories). Through SIL, IAD also maintains contacts with other international organizations that deal with the problems of quality and protection of water specifically, and of the environment in general. IAD usually organizes annual conferences or working sessions. At present IAD's research on the Danube and its tributaries is carried out in specialized groups created either according to scientific discipline (such as physics, chemistry, radiology, microbiology, and hygiene), or to topic (for example, dams, matter balance, and delta).

IAD also has a long history of publication. In 1967 it issued the monograph *Limnologie der Donau*, written by a group of prominent specialists, and a second part is forthcoming. A bibliography of works published since 1900 that deal with the Danube is in press. Hundreds of scientific and popular papers have also been put out, promoting knowledge about the river and its effects on the environment, and stimulating interest among specialists and economic and political authorities. This cooperative effort among the riparian and other countries has successfully helped to resolve various complex problems related to the Danube and provides a good example of effective international communication between Eastern and Western countries on the basis of mutual advantage and interest.

NOTE

This chapter is based on a paper presented at the conference "Environmental Problems and Policies in Eastern Europe," sponsored by the East European Program of the Woodrow Wilson International Center for Scholars, Washington, D.C., 15–16 June 1987. With the permission of the author, the editor has considerably shortened and edited the original paper, for purposes of stylistic clarity.

III
COUNTRY STUDIES

8

THE USE AND PROTECTION OF WATER RESOURCES IN BULGARIA

Georgi Gergov

Bulgaria covers 111,000 square kilometers of the northeastern part of the Balkan Peninsula and lies in the temperate geographical latitudes. The Danube in the north and the Black Sea in the east are its natural borders. The borders dividing Bulgaria from Yugoslavia in the west, Greece and Turkey in the south, and Romania in the north and northeast are, in contrast, a product of historical, ethnographic, and socioeconomic factors. In altitude the territory of Bulgaria varies from 0 meters above sea level at the seaside to 2,925 meters on the peak of Moussala in the Rila Mountains, with an average altitude of 470 meters (see table 8.1).

Climatic conditions in Bulgaria vary significantly and are determined by the mixed influences of the northern Atlantic Ocean, the Mediterranean Sea, Continental Europe, Siberia, North Africa, and the desert of Asia Minor. Continental influences prevail, however.[1] Generally, Bulgaria is characterized by a drought-prone climate. Rainfall is estimated to average 690 millimeters annually, which is similar to the precipitation in most southern and central European countries. Bulgaria's uniqueness is due to a high total evaporation rate of 520 millimeters. The average depth of the annual runoff is 170 millimeters, or only one-quarter of total rainfall. The maximum rainfall values measured in the mountains exceed 1,300 millimeters. The runoff coefficient varies between 10 and 30 percent and averages 26 percent for the whole country. It reaches 50 to 60 percent only in the highest mountains.

Table 8.1
HYPSOMETRY OF BULGARIA

Height above sea level (meters)	Area (percent)
0–200	31.5
200–500	34.5
500–1000	21.5
1000–2000	11.7
More than 2000	0.8

The high-altitude zones and the complex relief of Bulgaria are favorable to a dense river network with great longitudinal inclinations. The density of the river network varies greatly, from less than 0.1 kilometers per square kilometer in Dobrudja to 3 kilometers per square kilometer and more in the high mountains, with 0.176 kilometers per square kilometer the average for the whole country. According to accepted hydrological classifications, Bulgaria's inland rivers are of the small and average type. The largest, the Maritsa, has a water catchment area of 21,000 square kilometers, and the longest, the Iskur, is less than 370 kilometers long. The total length of Bulgaria's river network, including all perennial streams, is about 19,500 kilometers. The rivers in the mountains look like rapids or cascading river shoots in rocky river beds. The great longitudinal inclinations allow for high flow velocity, reaching 304 meters per second and more.

The great density of the river network and the high flow velocity are favorable for rapid runoff formation, and floods can occur after snows melt rapidly. (See table 8.2.) The high waters last from several hours to several days. Because these sudden floods in the mountainous areas of the country are extremely difficult to forecast, it is difficult to take necessary precautions to protect the population. The monthly distribution of the annual runoff of the Bulgarian rivers normally goes through three phases: the spring high water from March to June; the summer shallow-water period from July to October or November, when some rivers dry up completely; and the autumn-winter transitional period.[2]

The drought-prone climate of the country is manifested in intensive erosion as well as in the great load of suspended sediment.[3] During periods of high water, the concentration of suspended sediment is more than 50 kilograms per cubic meter. The rivers drag gravel, trees, and branches and have a great destructive force, overcoming natural or artificial obstacles such as bridges, engineering foundations, and residential buildings. Sad memories still linger of the catastrophic

Table 8.2
MAIN HYDROGRAPHIC CHARACTERISTICS OF SOME
BULGARIAN RIVERS

River	Runs into	Catchment area (square kilometers)	Length (kilometers)	Mean slope (percent)
Ogosta	Danube	3,157	144	11.4
Iskur	Danube	8,646	368	6.7
Vit	Danube	3,225	187	9.6
Osum	Danube	2,824	314	6.7
Yantra	Danube	7,862	286	4.6
Roussenski Lom	Danube	2,947	197	1.7
Kamchia	Black Sea	5,358	245	2.9
Tundja	White Sea	7,883	350	5.4
Maritsa	White Sea	21,084	322	7.3
Arda	White Sea	5,201	241	5.8
Mesta	White Sea	2,767	126	14.7
Struma	White Sea	10,797	290	7.3

SOURCE: K. Ivanov et al., *Hydrology of Bulgaria* (in Bulgarian) (Sofia: Nauka i Izkustvo, 1961), p. 314.

floods on the Maritsa River in 1957, the Rositsa in June 1959, some small rivers in the vicinity of the Lovech and Vratsa regions of the Balkan Range in the 1960s, the Lesnovska River in the Sofia Valley in 1972, and the rivers in the Bourgas region after 1980.

The concentration of naturally diluted chemical matter in rivers is highest in shallow water and lowest in the period of flood waves.[4] The high concentration of certain diluted matter endangers the ecosystem and has stimulated concern among hydrobiologists and environmentalists.

Bulgaria's total surface-water resources are estimated at about 20,109 cubic meters, not including the Danube's usable water resources.[5] Nationwide surface-water resources are very limited, and Bulgaria suffers from water shortages because of its drought-prone climate and the intensive use of water.

THE USE OF BULGARIA'S WATER RESOURCES

Bulgaria's limited water resources and the country's rapid industrialization call for intensive water use; at times, critical shortages have imposed restrictions on daily use of water. In 1985, water consumption reached 220 liters per day per capita. Irrigation, power production, industrial production, and communal services are the major consumers of water.

About one-third of Bulgaria's water resources are used for agricultural irrigation. Gravitational irrigation is most common (70 to 75 percent), with a 26,000-kilometer canal network. Large Bulgarian-fitted water-sprinkler systems have also been built, and experiments with drop irrigation are being conducted on a limited scale. A drainage system has been built for 90,000 hectares, and more than 130,000 hectares are protected from floods. More than 2,000 kilometers of river training works have been built so far, in addition to the 350 kilometers of dikes along the Danube started already in the 1920s. These and many other facilities were designed and built without the benefit of special research on the natural development of riverbed processes; in some cases, the experiences of other countries were influential. To increase further the availability of water, 40 large artificial lakes and more than 2,500 small artificial reservoirs have been built. The total volume of dam-filled waters amounts to about 33 percent of annual water resources. The recurrent filling of the reservoirs has greatly improved the regulation of surface-water runoff.

Water resources are regulated most effectively in the mountainous zone of southern Bulgaria. For example, in the Maritsa River basin, artificial lakes have a total volume exceeding 2 billion cubic meters, while in Dobrudja, the dry northeastern part of the country, the comparable figure is only 100 million cubic meters. Small water impoundments with a volume up to several million cubic meters have been built in the flat regions of the country. They are intended to meet the needs of local vegetable and fruit production and animal husbandry.

The construction of artificial lakes has been irregular. The majority of reservoirs (57 percent) were built between 1955 and 1960, and 25 percent between 1960 and 1965. Construction of water reservoirs in the last ten years has been considerably reduced, giving rise to anxiety among specialists. (Table 8.3 shows the distribution of artificial lakes per type of users.) The data suggest an insufficient number of all-purpose water reservoirs; that the recurrent use of accumulated waters is not well organized can be explained by the absence of a uniform economic approach to the construction of these reservoirs. A direct consequence has been a shortage of water for irrigation and industry in the last four or five years.

About 140,000 hectares of farmland in Bulgaria are irrigated with water from the Danube. In 1983, less than 8 percent of total water consumption (chiefly for irrigation) was met with water from the Danube. It is estimated that this figure can be raised to 20 percent and the acreage will double after the reservoirs on the Lower Danube are constructed. The government foresees a 40-percent increase in the area of irrigated land by the year 2000. Changes in modern agrotechniques should also reduce irrigation norms and increase the intensiveness of agricultural production in the agroclimatic region of the country.[6]

The production of hydropower accounts for a major share of water consumption in Bulgaria. In 1985 there were 88 hydroelectric power-generating stations with a total installed capacity of 2,000 megawatts, or one-third of the country's entire hydropotential. The amount of water used by the hydroelectric stations is supposed to drop from 33 percent of water resources in 1986 to 10 or 11 percent by the end of the century. According to data in the *Statistical Yearbook of 1986*, 5.4 percent of Bulgaria's total electric power is produced by hydroelectric stations.[7] The development of thermal and nuclear power generation in the last decades has reduced the share of hydroenergy in the total energy balance of the country, while the total amount of water-produced energy has increased sharply. Water can be saved for irrigation through increased reliance on nuclear power.

In recent years, Bulgaria has seen the construction of large hydro-power complexes in its mountainous regions—for example, along the Arda, Belmeken-Sestrimo, Petrohan, Vuch-Krichim, and Batak rivers. The largest and most complex project for national hydropower

Table 8.3
DISTRIBUTION OF BULGARIAN RESERVOIRS
AMONG USERS

Consumer	*Accumulated water volume (percent)*
Fish pool	0.08
Energy production	23.70
Equalizer	0.08
Portable water supply	5.63
Industrial water supply	10.81
Human needs	0.01
Complex purposes	21.72
Irrigation	37.97
Total	100.00

production is the Chaira pump-accumulation hydropower station, currently under construction. Thanks to hydroelectric power, peak electricity demands are now satisfied.

Large-scale redistribution of water resources has been an effect of the construction of hydropower complexes, especially in the mountainous regions of the south. Belt-strip water-catchment areas have been built at various altitudes to direct rapid and clear mountain streams northward toward the bigger consumers of water; in some cases, however, these catchment areas have led to the drying up of some high-mountain zones. A rapid change in local ecosystems resulted: forest and low-stem vegetation died, animal species migrated, riverbeds and dry valleys were buried in debris, and springs with drinking water dried up. Unfortunately, it would be very difficult and costly to reverse these changes, and programs to do so have not yet found their way into the national plans for environmental protection.

Another permanent source of demand is potable water for communal use. Eighty-one percent of Bulgaria's towns and villages, which are inhabited by 98.2 percent of its population, have a centralized water supply. Industry and recreational facilities are also important consumers. Industry uses 10 to 15 percent of the water resources, while potable water amounts to 7 to 8 percent.[8] Two-thirds of the water used by industry comes from underground sources.

In Bulgaria, the Danube's potential for hydropower generation remains practically unused. (The annual mean discharge is about 6,000 cubic meters per second.) Following the successful construction of a number of reservoirs for general use in the middle and upper stretches of the river, the construction of two reservoirs on the Lower Danube (the Bulgarian-Romanian stretch) is forthcoming. These reservoirs will allow for an expansion of irrigated fields on both sides of the river, increased power generation, and improved navigation. Navigation is particularly important, considering the extensive use of the river as a water route to the countries of Central and Eastern Europe. Construction of the two reservoirs will, however, have a somewhat unfavorable impact on the environment. The exact effects are not yet known. For example, irreversible riverbed changes in the Bulgarian-Romanian stretch of the river may take place. Concentrations of pollutants have already increased in some parts of the river.

In conclusion, water use is intensifying rapidly in Bulgaria. In some cases available water resources are nearly exhausted.

PROTECTION OF WATER RESOURCES IN BULGARIA

Environmental protection, particularly the protection of water resources, has for a long time been an important concern in the intel-
lectual and economic life of Bulgaria. In 1911 a group of scholars
appealed for environmental conservation. In 1920 the first law concerning water use and environmental protection was promulgated.

The protection of water resources in Bulgaria includes several elements: an assessment of the hydrochemical composition of natural
waters; adequate supply of water in the national economy; preservation of the reproductive capability of the major water sources; and
preservation of naturally formed water and dry-land biotopes. Until
recently there was no system for controlling the protection of water.
Consequently, pollution levels in parts of some rivers are inadmissibly
high, with negative consequences recognized by both specialists and
the public.[9] Frequently, downstream stretches of rivers have dried up
because runoff has been retained completely in the reservoirs, to be
redistributed among consumers. Thus long sections of rivers dry up
for long periods of time, while the riverbeds become overgrown with
vegetation and fill with debris from the adjacent banks.

Ecosystems and climate are changing. Examples can be found
throughout the country: the riverbeds downstream from the Pya-
suchnik and Kamchia reservoirs are densely overgrown with shrub-
bery, while the riverbed below the Iskur Reservoir is filled with debris
from the adjacent mountain slopes. Equally alarming is the widespread practice of building new embankment-shielded linear riverbeds that often lie outside the existing riverbeds. These linear riverbeds
channel rivers into new beds and sometimes lead them through new
geological conditions, while the old riverbed is abandoned or, during
construction, filled and leveled with the adjacent terrain. The new
riverbeds have no meadows, willow groves, or reefs. Fish die, the
hydroecosystem is destroyed, and the marshland and its water-loving
plants disappear.

All this happens in the name of protection. An example is the mouth
of the charming Kamchia River, which has been declared a protected
area. Beautiful sections of the lower reaches of the Lom, Struma,
Osum, Tsibritsa, and Topolnitsa rivers, which have been cut, have
dried up. Over time the rechanneled rivers begin to meander, threatening to destroy their new embankments. Examples of this can be
seen along the Struma River near the town of Radomir, in the lower
reaches of the Osum River, and on the Lesnovska River at the village

of Lesnovo. Occasionally the rivers threaten the stability and safety of railway transport, high-voltage transmission lines, and bridges.

The widespread extraction of sand and gravel from pits in riverbeds and from flooded banks has also endangered some rivers.[10] Some pits are several times deeper than the riverbed, while others stretch horizontally across miles of terrain. They are barren of soil and humus cover and are sparsely overgrown with thistles and weed. The deep pits drain the adjacent terrain, and underground water levels drop to several meters. Vegetation in some areas is doomed to die because of lasting droughts, notably along the Iskur, Struma, Chepelarska, Maritsa, Ogosta, and Stryama rivers, most of which have mines along them. Quarrying entails the degradation of riverbeds for long stretches, up to 10 to 20 kilometers upstream or downstream from a pit, while the impact on the environment often spreads over many miles.

These changes also damage, and sometimes destroy, structures built along the rivers. They have caused casualties or emergencies on the Struma, Maritsa, Topolnitsa, and Stryama rivers. Highway and railway bridges have been destroyed on the Chepelarska, Eleshnitsa, Topolnitsa, and Stryama rivers. The old bridge in the center of Plovdiv on the Maritsa River collapsed some years ago. The bridges over the Struma River in the region of Blagoevgrad are also threatened. The construction of a bottom threshold in the lower stretch of the 500-year-old historical bridge on the Maritsa River in Svilengrad has saved it from a similar fate. This probably is the reason for the destruction of the left-side support of the famous bridge over the Yantra River in the town of Byala, the work of the master builder Kolyo Ficheto. Long-distance power transmission lines and pipelines have also been harmed; embankments and dikes have been swept away, causing costly damages.

The purity of natural waters is greatly compromised. The Struma, Iskur, Maritsa, Yantra, Topolnitsa, and other rivers are polluted for long stretches, and they have already lost their properties as natural water receptacles. They are highly polluted with organic matter, oil products, toxic matter, and plant-control chemicals. There have also been cases of flash pollution of technological origin or traceable to cattle-breeding farms.

The atmosphere in several urban-industrial regions in Bulgaria—notably Sofia, Kremikovtsi, Pernik, Razlog, Plovdiv, Pazardjik, and Dimitrovgrad—has been contaminated by gases and high dust concentration, both of which are harmful to living organisms. Local pollution, combined with long-distance pollutant transfer, has produced acid rain, which contaminates the natural waters with sulfur and nitrogen compounds and limits the usability of river waters. Furthermore, these effects of atmospheric pollution reduce the expected effectiveness of purifying plants and of closed-cycle technological systems.[11]

Along the Bulgarian Black Sea coast, piers, fortified walls, coastal installations, moats, and bottom threshold installations have been built on a mass scale to protect the coast, beaches, and ports from the harmful impact of sea waves. Frequently, however, engineers are surprised by the negative effects of these environmental protection measures on the adjacent terrain. The activated abrasive processes can even destroy the installations themselves.

The absence until recently of an effective water protection program has resulted in the steady growth of pollution. The major government agency responsible for controlling the pollution of water resources in Bulgaria is the State Committee for Environmental Protection (SCEP). Thanks to its efforts, some frequent violators have been subject to sanctions, including the closing down of their facilities. A number of polluting industries have been moved out of Sofia in compliance with the recommendations of SCEP.

SCEP has created, and regularly updates, a list of water users and water polluters. It also publishes bulletins providing evaluations of the environment, supplemented by maps of water contamination. This information has helped to increase public awareness of pollution. In its environmental programs, Bulgarian television has actively followed and condemned polluters. Nationwide plans aimed at protecting water resources and assuring their optimal use have also been developed. Numerous laws provide the legal basis for Bulgaria's water protection program.

To ensure implementation of these legislative measures, about 2.5 percent of the national budget was allocated for pollution-related expenditures in 1986. This amount covered capital installations, construction, assembly work, and other activities pertaining to erosion control, recultivation of harmed terrains, biological and plant control, and afforestation. According to data of the Central Statistical Office,

> In 1986 new purifying plants were built with a capacity to capture more than 19,000 tons of harmful matter contaminating the atmosphere and to purify over 40 million m^3 of waste water a year. An area of 638,000 decares has been drained, and 16,500 decares of saline and harmfully acidic soils have been chemically ameliorated in order to restore fertility. An area of 8.3 million decares has been treated biologically and in-

tegrated plant control methods are being used on it. Over 32,000 decares have been afforested and 122 new projects have been announced under the Nature Protection Act. Waste-free and low-waste technologies introduced during the year will assure the utilization of 950,000 tons of waste products per year.[12]

It is envisaged that by 1990 the proportion of purified wastewaters will reach 63 percent of the total water mass.

Some achievements from the past several years can be cited: the purification of the Struma River following the construction of the Pchelina Reservoir; the development of the Iskur hydrocomplex in the Sofia Valley; proclamation of vast territories as protected areas (for instance, the Pirin Mountain, with an area of more than 27,000 hectares); construction of large purifying plants in Steneto, Longoza, and Parangalitsa; and the general program for construction on the Bulgarian seaside.

In 1986 a new "general scheme for use of water resources in Bulgaria" for the period up to the year 2020 was developed, including a scientifically based estimate of water resources and a plan for their distribution. New norms of water use, improved standards for water protection, and recycling in technological processes underlie the general scheme.

MONITORING AND RESEARCH

Although the Hydrographic Service had been founded in 1935, the year 1950 saw fresh impetus given to the development of the hydrometric network in Bulgaria. In that year, the Chief Hydrological and Meteorological Office (CHMO) was placed under the Bulgarian Academy of Sciences. The national hydrometric network at present comprises 272 stations (average density 440.4 square kilometers per gauge), which carry out daily measurements of water levels and temperatures and regularly measure water discharges. At 130 of these stations, samples have been collected regularly since 1950 to determine the suspended sediment concentration in river water.[13] For more than twenty years now,[14] sixty-one stations have regularly collected samples to study the suspended sediment grain size distribution. Over more than thirty years, 148 hydrometric stations and a number of auxiliary points, covering practically the entire hydrographic network in Bulgaria, have also been measuring the concentration of the chemically diluted matter in river water. These investigations extend to many lakes in the high mountains, artificial lakes, and the Black Sea coast. Regular sampling also occurs along the Danube's Bulgarian-Romanian stretch. The hydrogeological network also comprises 1,256 wells, springs, and bores; hydrochemical analyses are carried out regularly in two hundred of them.

SCEP has a gauge network of its own for environmental monitoring. Automatic devices have been installed in the most contaminated regions. SCEP uses both aircraft and boats to monitor the purity of the seaside.

Experimental stations have been built to conduct research at key points around the country. Their investigations have examined karst hydrology in the Nastan-Trigrad region; the underground water regime and water deterioration in the irrigation fields around Plovdiv; the effect of forest on runoff formations in the Parangalitsa Mountain Wildlife Refuge; the state of the Danube near the town of Nikopol; and the lithodynamic processes in the coastal zone of the western Black Sea near the mouth of the Kamchia River. In southern Bulgaria a basin is being constructed to investigate the anthropogenic impact on river runoff, ameliorative hydrology, and urbanistic hydrology. This basin covers a vast territory including several rivers, part of an irrigation field, and many hydrometric and rain gauges.

In 1986 a program was established for the comprehensive study of a large artificial lake in central Bulgaria, involving simultaneous climatic and microclimatic measurements at a floating base on the lake and at many points adjacent to the lake. Instruments measure air and water pollution, water and sediment inflow and outflow, currents, and sediment deposits in the lake. Measurements and processing of information will be expanded by means of an automated telemetric system of Bulgarian-made electronic devices and microcomputers.

Video equipment is also being tested for underwater measurement and registration.[15] Small pilotless aircraft have been developed for the remote sensing of the spread of pollutants in lakes and water reservoirs, and for an assessment of the condition of agricultural and forest plantings. The information collected in the national hydrometric gauge network is processed and catalogued in the CHMO Cadastre Center.[16] The center is being increasingly equipped with computers and Bulgarian-made microcomputers. Hydrological annuals and reference books summarizing the collected information are regularly published.[17] Reference books on the Danube are exchanged with the other countries in the Danubian water catchment basin.

Bulgaria participates in a number of international monitoring and control programs. The International Commission for the Danube, based in Budapest, supervises large engineering projects that are liable to obstruct international shipping or cause floods. The commission assures the maintenance of navigable conditions and the protection of the Danube's waters from chemical and radioactive pollution. Bulgaria also participates in the international research programs of the United Nations Educational, Scientific, and Cultural Organization; the World Meteorological Organization; the International Hydrology Program; and the Council for Mutual Economic Assistance.

Within Bulgaria, the supreme body coordinating and directing the use and protection of water resources is the State Waters Council, which works together with the Council of Ministers. Its prerogatives as an executive body of state power are great. Legislative power is in the hands of the Environmental Protection Committee of the Bulgarian National Assembly. Research programs relating to water protection are coordinated by the Bulgarian Academy of Sciences.

The Institute of Hydrology and Meteorology, the Institute of Water Problems, the Hydrotechnology and Amelioration Institute, the Higher Institute for Architecture and Civil Engineering, and the Oceanological Institute are all engaged in scientific research pertaining to the use of water, while design organizations such as Vodproekt, Energoproekt, and Vodokanalproekt are charged with the implementation of water-related development and engineering projects. Scientific and practical problems of environmental control are handled by SCEP in the Council of Ministers and by the Hygiene and Epidemiology Institute of the Ministry of Public Health. Some specialists admit that coordination of the activities of these organizations is still inefficient and insufficient.

PROBLEMS OF WATER RESOURCE POLICY IN BULGARIA

The discovery and management of new water sources holds top priority in water resource policy. The Danube is a great reserve, as are underground water deposits.[18] Still unassessed is the potential for the direct use of sea water. The possibilities for exploiting water resources in the mountainous regions have not yet been adequately explored. Research into the artificial stimulation of rainfall has recently begun in Bulgaria.

Special importance is also attached to the distribution of water, regulation and reduction of water consumption, and improved water purification. Necessary measures include installing more purifying equipment, curbing the use of artificial fertilizers and chemical preparations for pest control, pursuing the introduction of waste-free technologies and closed-cycle industrial water supply systems, limiting the pollution of the natural environment with heavy metals, and implementing strict measures to keep underground water pure. The capacity of purifying plants to treat all contaminated water is sometimes insufficient. Some big polluters continue to put off the construction of purification installations, thereby reducing the effect of the efforts of other enterprises.

The automation of hydrometric measurements and observations is also very important. Complete automation will be difficult, however, because at present several organizations are in charge of it, including CHMO, SCEP, the Artificial Lakes and Cascades Office, and the Enterprise of Water Economy. The creation of a national information base has been inadequately coordinated.

Modern computer equipment allows for the full automation of the transfer, storage, treatment, and retrieval of hydrological information, but so far these possibilities have not been fully exploited.[19] Hydrological forecasts are especially important; their use will make it possible to exploit the water in artificial lakes more effectively and reduce damage caused by catastrophic floods. Recent research shows that the pollution of deep bodies of water (for example, on the Bulgarian coast) can be successfully controlled by being treated with multichannel space images in the visible and nearly visible infrared part of the spectrum, or with analogical aerial photographs and ground measurements of the brightness of reflected sunlight.[20]

The full impact of artificial lakes needs to be assessed. Especially important is the effect of these lakes on the development of natural processes in adjacent zones, irrespective of the targets and motives of the construction of the lakes. The proper location of engineering installations in or near rivers is also an important consideration. Engineers must be familiar with the regularities of the fluvial processes and river forms near human-made facilities.[21] Underground waters must be more effectively protected from pollution.[22] The effects of urbanization on conditions for the formation of surface runoff and the quality of water also deserve more extensive study.

CONCLUSION

In conclusion, the task facing hydrotechnical and water economy specialists in Bulgaria is to assure the effective use of water resources in order to realize national economic goals while minimizing the negative effects on the environment. A successful approach will require

- restricting water consumption in technological cycles in accordance with actual needs,
- implementing closed-cycle water supply systems including the recycling of water in industry,
- introducing new methods of irrigation with reduced water consumption,
- developing rational engineering schemes for hydrotechnical and hydroenergy construction in the river basins,
- curbing water pollution,
- restricting river mining, and
- improving the system of price setting for water use.

Educational programs are especially important. We all should learn to love and protect the natural environment in which we work and live, if only for our own sake. Modern theoretical models and computers make it possible to realize environmental goals, and we should direct our joint efforts toward training young engineers to take advantage of these possibilities. Nature and life are complex and eternal. Protecting them is a great responsibility for present and future generations—and we should live up to it by extending and deepening education about environmental protection.

NOTES

This chapter is based on a paper presented at the conference "Environmental Problems and Policies in Eastern Europe," sponsored by the East European Program of the Woodrow Wilson International Center for Scholars, Washington, D.C., 15–16 June 1987. With the permission of the author, the editor has considerably shortened and edited the original paper for purposes of stylistic clarity.

1. L. Sabev and S. Stanev, "Climatological Regions in Bulgaria and Their Climate," *Bulletin of the Institute for Hydrology and Meteorology* 5 (Sofia: 1959).
2. T. Panayotov, "Unsteadiness of the Annual Distribution of Stream Flow Hydrological Phases and Hydrological Seasons of the Rivers in Bulgaria" (in Bulgarian), *Bulletin of the Institute for Hydrology and Meteorology* 20 (Sofia: 1972): 82–100.
3. R. Papazov, *Sediment Transport Rate of Open Channel Flow* (in Bulgarian) (Sofia: Bulgarian Academy of Sciences, 1981); and D. Pechinov, "Some Questions on the Formation and Regimentation of Suspended Load in the Rivers of Bulgaria" (in Bulgarian), *Bulletin of the Institute for Hydrology and Meteorology* 17 (Sofia: 1970): 1–56.
4. T. Tzachev et al., *The River Water Organic Matter Pollution in Bulgaria* (in Bulgarian) (Sofia: Bulgarian Academy of Sciences, 1977), p. 143.
5. K. Ivanov, "Hydrofacies of the Inland Waters in Bulgaria" (in Bulgarian), *Bulletin of the Institute of Hydrology and Meteorology* 12 (Sofia, 1962).
6. E. Hershkovich, "Agroclimatic Division of Bulgaria into Regions" (in Bulgarian), *Bulletin of the Institute for Hydrology and Meteorology* 17 (1970): 203–27.
7. *Statistical Yearbook of Bulgaria* (in Bulgarian) (Sofia: Central Statistical Office, 1986), p. 681.
8. Ibid., p. 681.

9. G. Gergov and I. Nenov, "Alert about the Bulgarian Rivers" (in Bulgarian), *National Front Newspaper*, no. 12088 (1985), p. 2.
10. J. Jordanov, "The Yield and the Resources of the Natural Building Materials in Bulgaria" (in Bulgarian), *Technical Report on the Institute of Building Management* (Sofia: 1976). See also D. Pechinov, G. Gergov, and I. Nenov, "Some Hydrological Aspects of Sand and Gravel Yield from Riverbeds" (in Bulgarian), *Hydrology and Meteorology* 2 (Sofia, 1979).
11. J. Jordanov, "The Yield and the Resources."
12. *Rabotnichesko Delo*, no. 30, 30 January 1987.
13. A. Petkov and D. Pechinov, "Suspended Sediments in the Bulgarian Rivers" (in Bulgarian), *Bulletin of the Institute for Hydrology and Meteorology* 1 (1958).
14. G. Gergov and S. Blyskova, "Grain Size Structure of Suspended Sediment in Bulgarian Tributaries of the Danube" (in Russian), *Proceedings of the XIII Conference of the Danube Countries on Hydrological Forecasting* (Belgrade, 1986), pp. 217–24.
15. S. Kuprianov et al., "Methods and Means of Obtaining and Treating Primary Visual Information in Maritime Ecological and Biophysical Investigation" (in Bulgarian), *Technical Report of the Central Laboratory on Management Systems* (Sofia, 1985).
16. D. Mandadjiev and T. Bojkova, "The Contents and Structure of the Data Bank at the Automated Information System on Hydrology and Hydrogeology" (in Bulgarian), *Problems of Meteorology and Hydrology* 6 (Sofia, 1987).
17. *Hydrology: Handbook of Bulgarian Rivers*, vols. 2–5 (in Bulgarian) (Sofia).
18. P. Betzinski, "On Underground Water Resources in Bulgaria" (in Bulgarian), *Hydrology and Meteorology*, no. 4 (Sofia, 1976), pp. 32–39.
19. S. Gerasimov and B. Semenov, eds., *Automated Treatment of Hydrological Information on the Hydrological River Regime* (in Russian) (Moscow: Gidrometeoizdat, 1987).
20. G. Gergov, "Application of Distant Methods in Hydrology" (in Bulgarian), *Journal of the Bulgarian Academy of Sciences*, no. 4 (Sofia, 1985): 51–56.
21. G. Gergov and I. Nenov, "Alert about the Bulgarian Rivers"; and G. Gergov, "The Fluvial Bed Processes in Bulgaria" (in Russian), *Proceedings of the Institute for Hydrology* 216 (Leningrad, 1974): 36–51.
22. C. Antonov and B. Raikova, "An Attempt at Typification of Fresh Underground Waters in Bulgaria According to Stages of Vulnerability to Pollution" (in Bulgarian), *Hydrology and Meteorology*, no. 6 (Sofia, 1978): 11–20; and B. Raikova and E. Stoeva, "Quality Control and Protection of Underground Waters in Bulgaria" (in Russian), *Proceedings of the International Symposium on Groundwater Monitoring and Management* (Dresden, 1987).

9

"THE FUTURE HAS ALREADY BEGUN": ENVIRONMENTAL DAMAGE AND PROTECTION IN THE GDR

Joan DeBardeleben

On ten days during January and February 1987, the authorities in West Berlin issued "smog warnings" or "smog alarms."[1] In the surrounding German Democratic Republic (GDR), people also learned of the extremely high levels of air pollution by listening to broadcasts from West Berlin, while the authorities remained silent, explaining the unusual atmospheric conditions with forecasts of foggy weather. On those days, an estimated 80 percent of the sulfur dioxide, a major component of West Berlin's smog, originated in the GDR. (Under normal conditions West Berlin "imports" an estimated 45 percent of its sulfur dioxide pollution and two-thirds of its dust pollution.)[2] On some days in Berlin one can smell the bad air, although on other smoggy days the air seems surprisingly fresh, because some harmful pollutants are difficult to detect without appropriate air quality control equipment. In the highly industrialized southern regions of the GDR, residents know that the air is polluted: they see the dust on their laundry and smell it in the air.

Only by arduous scrutiny of GDR scientific publications can one begin to garner enough information to make a rough assessment of pollution levels in East Germany. The popular press assures the public about continued progress in overcoming the few deficiencies. On 16 November 1982, the Council of Ministers adopted an official "Order for Ensuring the Secrecy (*Geheimschutz*) of Environmental Data."[3] The order itself was never publicized, but its effects on public information policy are obvious to any concerned citizen. Virtually no information is published on actual pollution levels, except for occasional indirect references in scientific sources.

Despite official reassurance, public awareness is growing in the GDR, although its extent is only a matter of speculation. The nuclear accident at Chernobyl in April 1986 gave an additional impetus to the growing popular concern. By 1986 some sixty thousand citizens of the GDR were members of the officially sponsored Society for Nature and the Environment and at least several hundred were active in the more challenging unofficial environmental groups connected to the Evangelical church. In recent years these unofficial groups have held "environmental worship services" in heavily polluted areas, such as the small town of Möblis some twenty kilometers south of Leipzig.[4] The motto "In Möblis the future has already begun" provides a warning—if more is not done now, the future will look as gray and smell as foul as towns like Möblis do now.

THE PROBLEMS

The citizens of East Germany suffer from the effects of numerous kinds of environmental damage, but controls on public information prevent them from knowing its dimensions and implications. Although people complain about the bad air, environmental deterioration may not be a major source of discontent because its effects and scope are often hidden. Nonetheless, the hazards to health and well-being are very real.

Brown Coal

The major source of air pollution and the most serious and pervasive environmental problem in the GDR is the burning of domestic brown coal for heat and energy. The coal, while varying from area to area, has a high sulfur and ash content and therefore produces high sulfur dioxide and dust emissions.

Brown coal is the GDR's only significant energy resource other than uranium ore that can be processed for use as fuel for nuclear reactors. In the early 1980s the GDR's leaders increased reliance on domestic brown coal as the primary source of energy, reversing the 1970s shift to imported oil. Confronted by rising foreign debt and oil prices in the late 1970s, import of fuel seemed increasingly untenable. Beginning in 1975, imports of Soviet oil above the negotiated annual quotas had to be financed with hard currency, and in 1982 these quotas were reduced by about 10 percent. Since that time Soviet oil imports have stabilized at around 17 million tons and gas imports at just over 6 million tons annually. Much of this oil is, however, used in the petrochemical industry and not as fuel. By 1985, 82.7 percent of electrical energy production was based on brown coal.[5] The increased reliance on this fuel has helped to lower imports and thus the GDR's foreign debt, but at the same time it has intensified air pollution. Emissions from large power plants are accompanied by smoke from residential coal ovens and from small outdated furnaces supplying individual factories. At the end of 1982, about 4.8 million apartments were still heated by coal ovens.[6] Western experts estimate that total sulfur dioxide emissions for the GDR in 1985 were about 5.8 million tons, the highest in Europe (as compared, for example, to about 3 million tons for West Germany, which has more than three times the population and more than twice the land area of the GDR).[7] This pollutant contributes both to smog and to acid rain, although the latter often falls far from its source, affecting neighboring countries more than the GDR itself. High smokestacks (200 to 300 meters) reduce intense pollution in the immediate vicinity but distribute it over a wider area. Although the GDR's sulfur dioxide pollution is high, some other pollutants, most notably nitrogen oxides, are emitted in the GDR in much lower quantities than in the Federal Republic of Germany (FRG).[8]

The air in the GDR's industrial south is the worst. Half of the sulfur dioxide emissions from the large brown-coal energy plants come from the district of Cottbus.[9] Cities such as Leipzig, Bitterfeld, Karl-Marx-Stadt, and Halle are also highly polluted, because the chemical industry, concentrated in the area around Halle, adds to the problem. In 1970 one East German scientist estimated that in these heavily polluted areas a reduction of 50 percent in emissions could extend life expectancy by three to four years, reduce heart and circulatory disease by 10 to 15 percent, and bring a long-term decline of 25 percent in respiratory malignancies.[10] Life expectancy in the Halle area is reportedly five years below, and the incidence of heart and respiratory disease 10 to 15 percent above, the levels elsewhere in the GDR.[11] Some East German scientists acknowledge that increasing cancer rates in the Cottbus area are at least in part due to high levels of air pollution.[12]

Human health is not the only victim of air pollution. An émigré expert estimates that some 90 percent of the GDR's forests are damaged and about 0.5 million hectares (15 to 20 percent) of the forests are already dead or dying.[13] The Erzgebirge, near the Czechoslovak border, leave a horrifying impression on visitors and local residents alike. Air pollution from neighboring Czechoslovakia, Poland, and West Germany contributes to the damage. East Germany's policy emphasizes planting "smoke-resistant" varieties of trees (maple, beech, birch, aspen, mountain ash) to replace the more sensitive spruce and pine.

Research into technological processes for neutralizing harmful pollutants emitted by power plants is advanced in the GDR, but practice is less satisfactory. For example, the lime-additive process, often mentioned in writings from the GDR, involves binding sulfur with lime but generally results in only a 30-percent reduction in sulfur dioxide emissions. High expenditures are required to transport the very large quantities of lime and waste ash, so that only selected and intermittent use of this method is deemed feasible in heavily populated areas or on extremely polluted days, for example. So far, this technology is operating on a trial basis at Vockerode, a 385-megawatt plant in the Halle region, but it will soon be introduced at three additional plants.[14] Import of expensive desulfurization equipment from Western firms requires spending valuable hard currency, and its effectiveness in GDR plants has not been proved. As a test project, the British have loaned the GDR some forty million pounds to finance import of smoke-desulfurization equipment (applying the Wellman-Lord process) for the relatively small Rummelsburg power plant (180 megawatts) in Berlin.[15] Existing equipment designed to remove dust particles from emissions is not always cleaned regularly, impairing its effectiveness.

Despite these difficulties, at the 1985 meeting on the environment of the United Nations Economic Commission for Europe the GDR joined other nations in an agreement to reduce its sulfur dioxide emissions by 30 percent by the year 1993, in relation to 1980 levels. Some Western experts question the economic feasibility of this goal, especially because no apparent progress had been made by the end of 1986. GDR spokespersons point to more efficient energy use, improved desulfurization of emissions, increased reliance on nuclear energy, and expanded use of centralized heating stations as the main methods to realize the commitment.[16]

Even before it is burned, brown coal leaves a permanent mark on the landscape and on human lives. Mining it is becoming increasingly expensive because reserves lie deeper in the ground and underneath established settlements. Between 1960 and 1980 at least seventy villages or parts of villages were sacrificed, and 190 kilometers of road (as well as sixty kilometers of waterway) had to be moved to accommodate brown-coal strip-mining.[17] Several other villages, especially in the Cottbus region and south of Leipzig, face the same fate in the future, as mining is to increase from about 260 million tons in 1980 to a planned level of 335 million tons in 1990. The residents are resettled as tons of groundwater and soil must be brought to the surface to allow access to the coal. Government spokespersons claim that recultivation means a positive transformation of the landscape in the affected areas through development of new lakes and recreational areas, along with an improvement in soil conditions through careful attention to the character and quality of the soil that is being replaced. In most cases this promise still awaits fulfillment.

The economic and ecological costs are enormous: agricultural land must be removed from cultivation; groundwater levels fall; water becomes polluted with sulfates, phenols, sulfuric acid, and particulate matter;[18] animal and plant habitats are disrupted; trees are felled; and new housing, schools, and other facilities must be provided for the resettled population. With the GDR already suffering from a shortage of housing and labor, these demands strain even further the limited resources available for improving the sociocultural environment. Nonetheless, GDR economists believe that continued exploitation of domestic brown coal is still the most economical alternative.

A Nuclear Future?

In the long run, the GDR plans to turn to nuclear energy. At present only some 11 percent of that country's electrical energy comes from nuclear plants, and it has only one large reactor complex: Griefswald (with four reactors, each 440 megawatts), located near the northern sea town of Lubmin. In addition, a small seventy-megawatt plant, also used for research and education, is located about thirty kilometers north of Berlin near Rheinsberg. By 1990, 15 percent of electrical power should come from nuclear energy, and long-term projections set a target of 54 percent by 2020, as brown-coal reserves approach depletion.[19] Immediate plans for expansion include a doubling of capacity at Griefswald and the completion of a new complex near Stendal, in the Magdeburg region, about a hundred kilometers west of Berlin and less than fifty kilometers from the West German border. The first reactor (a thousand megawatts) is to begin trial operation by 1990.

Like the other East European countries, the GDR uses Soviet-designed pressurized-water reactors, not the Chernobyl-type channel graphite-moderated reactor. Major components for the newer generation of pressurized-water reactors in the GDR will be produced at the Skoda works in Czechoslovakia. Despite reports of a small accident at the Griefswald plant in 1976,[20] the East German safety record thus far has been good. Existing reactors are, however, not outfitted with the concrete containment structures characteristic of most pressurized-water reactors in the West. But the new ones probably will be. The East German government may find it necessary to reconsider its nuclear commitment in view of the high investment costs associated with the development of nuclear power. Economic—and not ecological—considerations probably explain the consistent lag in fulfillment of plans in the past ten years.[21] Since the Chernobyl accident, however, there is increasing evidence that the GDR's leadership may be reevaluating the scope of its commitment to nuclear power, and that certain aspects of reactor safety (including cooperation with the West German nuclear industry) are also being studied.[22]

The East German citizen hears little from the domestic media about the possible hazards of nuclear power. But access to West German broadcasts assures most East Germans the full gamut of information. An exception was the period from 29 April to 6 May 1986, when *Neues Deutschland*, the largest daily, published several articles about the Chernobyl accident.[23] Radiation levels in the GDR on the peak days of 30 April and 1 May were printed on page one of the newspaper but in a form of little use to the average citizen.[24] The media offered reassurances that human health was not endangered at any time, also citing West German sources. On 5 and 6 May additional articles in *Neues Deutschland* cited past nuclear accidents in the United States. Although these reports may have reassured the population that Soviet plants are little worse than Western ones, they might also have raised doubts about the overall safety of nuclear technology—if even Western technology is unsafe. To counter such doubts, East German sources subtly hinted that GDR plants are more reliable than their Soviet counterparts.

It seems doubtful whether any milk or other food products were withheld from sale immediately following the accident, as was the case in some neighboring countries (including West Germany) when radiation levels were presumably similar in at least some parts of the GDR. East German shoppers were reportedly glad to buy fresh Polish vegetables exported to the GDR after being rejected by the FRG because of their high levels of contamination. One Western report indicates that some dry feed was withheld from use for dairy and slaughter cattle early in 1987.[25] No information has been made available to the GDR citizen, however, about post-Chernobyl radiation levels in food, whereas some newspapers in West Berlin continued to publish measurements daily more than a year after the accident. GDR citizens were able to watch a Soviet film on the accident that was aired on GDR television just a few days after its debut on Soviet television on 25 February 1987. The film, called "The Warning," emphasized how unthinkable nuclear war is, given the catastrophic effects of Chernobyl. On 8 February 1987, apparently for the first time, experts from the GDR State Office for Atomic Safety and Radiation Protection met with about a hundred "environmentalists" (it is not clear who was included in this group) and members of a mass women's organization (*Demokratische Frauenbund*) to allow airing of questions about the safety of nuclear power.[26] What the effect of a more open information policy would be is difficult to know. The hypothetical and completely invisible dangers of nuclear power may well seem worth the risk if nuclear power can assure an improved standard of living. Furthermore, these dangers may seem preferable to the demonstrable disadvantages of a brown-coal economy. The average citizen of the GDR may well affirm Erich Honecker's sentiment stated at the Eleventh Socialist Unity Party (SED) Congress in April 1986, just before the Chernobyl accident: nuclear power should be expanded, "not least for the sake of the environment."[27]

Land, Food, and Water

Agriculture is one of the worst sources of environmental damage in the GDR, while the ecological consciousness of agricultural practitioners lags considerably behind that of industrial managers. Fertilizers and pesticides continue to be employed extensively to increase crop yields, despite a slight reduction in the early 1980s compared to the upward trend in the 1960s and 1970s. (See table 9.1.)[28]

Some of these pesticides contain DDT and mercury compounds. Although application of DDT is gradually declining, it continues to be available to the average citizen for use in private gardens and is also a component in widely available wood preservatives. Disturbing concentrations of DDT have been measured by GDR scientists in mothers' milk, birds, and plants. High levels of mercury have been found in some birds and fish. Similarly, excessively high (and in some cases rising) nitrate levels make water unsafe for drinking in some areas, especially the countryside. Nitrate consumption through food and water in the GDR almost doubled between 1970 (78 milligrams per capita per year) and 1983 (150 milligrams per capita per year).[29] In some cases, scientists have advised mothers to nurse their infants or to use bottled mineral water to feed them. Levels of these and

Table 9.1
TOTAL TONS OF CROP-PROTECTION AGENTS SUPPLIED TO EAST GERMAN AGRICULTURE, 1965–85

| | Tons | |
| | Active substances (includes herbicides) | |
Year		Herbicides
1965	8,219	6,197
1970	18,567	13,758
1971	18,037	13,979
1972	21,901	15,346
1973	21,957	12,150
1974	22,090	14,694
1975	22,480	15,004
1976	23,665	16,243
1977	24,502	16,915
1978	25,298	17,080
1979	26,715	17,999
1980	27,009	18,067
1981	26,951	19,277
1982	26,744	19,432
1983	25,951	18,773
1984	25,985	18,758
1985	26,731	18,179

SOURCE: *Statistisches Jahrbuch der DDR* (Berlin: Staatsverlag der DDR, 1986), p. 191.

other contaminants in food and water supplies are not generally made public, but we may presume that they are significant enough to pose some health hazard, particularly for young children.[30] Average citizens know nothing of these dangers, even when produce from their own small gardens is affected.

Agricultural chemicals are increasingly sprayed from airplanes, not always with appropriate care given to wind conditions; as a result, rivers, streams, and nearby livestock may be inadvertently contaminated. (See table 9.2.) Local residents, unaware of the spraying, pick contaminated berries and mushrooms. Several cases of human illness and livestock deaths have been reported through unofficial channels.[31] Fertilizer is often unloaded unpacked, resulting in heavy dust pollution, contamination of soil and water, and chemical burns on vegetation.[32] Several plant and animal species are now extinct or threatened with extinction in the GDR, largely as a result of the heavy use of chemical fertilizers and pesticides, as well as of the extensive land drainage programs of the 1960s and 1970s.[33] Although most of

Table 9.2
MINERAL FERTILIZERS SUPPLIED TO EAST GERMAN AGRICULTURE, 1950–1985

Year	Fertilizers (kilograms per hectare of agricultural area)			
	Nitrogen	*Phosphate*	*Potash*	*Lime*
1950	28.7	15.4	59.7	84.5
1955	32.4	20.5	65.6	109.6
1960	36.7	34.0	77.4	121.0
1965	65.1	49.3	92.1	212.2
1970	81.3	65.2	97.7	186.8
1975	107.7	70.1	112.2	206.3
1976	120.0	67.5	99.1	202.6
1977	122.5	67.9	75.0	172.5
1978	124.8	68.6	72.4	170.0
1979	119.0	66.3	87.3	151.1
1980	119.9	62.0	79.2	197.8
1981	119.7	59.8	96.0	198.3
1982	97.1	48.6	79.3	192.7
1983	111.0	53.3	67.9	222.4
1984	111.6	51.1	88.1	221.6
1985	123.7	51.2	88.4	228.8

SOURCE: *Statistisches Jahrbuch der DDR* (Berlin: Staatsverlag der DDR, 1986), p. 191.

these problems are not unique to the GDR, the absence of open debate there makes solutions more difficult. Nature protection zones have been expanded, however, and in 1985, 856 animal species and 136 plant species were under special protection.[34]

Because the GDR has a limited natural water endowment, contamination of groundwater with nitrates and other chemical pollutants is especially serious. In many areas water must be reused many times by various industrial and agricultural producers. Enterprises are encouraged to introduce technological processes that reduce water waste and improve purification to allow recycling in the production process. Much of the groundwater currently requires purification before drinking. In 1970, some 80 percent of the surface water was already classified as polluted or heavily polluted, unsuitable for fishing, sports, or drinking without treatment.[35] Many rivers (notably the Elbe, Pleisse, Salle, and Werra) are heavily polluted from the chemical, textile, paper, metallurgical, and potash industries, especially in the industrial south, and reports of major fish kills occasionally reach the West. Despite these problems, in the past few years the situation appears to have stabilized and in some cases improved.

Waste and Recycling

The GDR has been remarkably ambitious and quite successful with its policies on reuse of waste products as "secondary raw materials." Enterprises are required to include measures in their economic plans and, in accordance with *Verursacher-Verantwortung* (responsibility of the producer), are sometimes even expected to organize the recycling of worn-out consumer items they produce. For example, the VEB Petrochemical Combine Schwedt is responsible for recycling used motor and industrial oil, and the VEB Tire Combine Fürstenwalde recycles used tires.[36]

The Combine Secondary Raw Material Recovery (*Sekundärrohstofferfassung*) involves the public in recycling by establishing collection centers where citizens can return waste for payment. For example, 0.30 mark is paid for a kilogram of newspapers, 0.20–0.30 mark for recycled glass containers, and 0.50 mark for each kilogram of old textiles.[37] The Free German Youth also organizes voluntary collections. Large bins for old plastic containers can be found in supermarkets, and table waste for pigs is collected in cans on the streets. In 1980 some 916,000 tons of kitchen waste were collected from the population. Nonreusable glass bottles no longer exist in the GDR. Some 50 percent of paper production is based on old paper; scrap metal provides about 75 percent of the raw material for iron and steel production; scrap provides 35 percent of the raw material for copper, 40 percent for lead, and 20 percent for aluminum production. The reuse of industrial waste for secondary raw materials increased from twelve million tons in 1975 to thirty million tons in 1983.[38] These achievements are important to the GDR leadership primarily as economic savings, and ecological motives may be secondary. Nonetheless, the environmental benefits are significant.

An enterprise must receive authorization from the responsible local organ before it deposits waste products; authorization follows only after the enterprise provides sufficient evidence that the waste cannot be used as secondary raw material. Then the local authority approves disposing of the material at a specified dump.[39] Nonetheless, unregulated dumps are still common in the countryside, with private citizens often the worst offenders. Storage of toxic waste occurs at specially regulated sites, and so far there have been no reported violations. The GDR also accepts waste from the FRG and West Berlin in exchange for hard currency. Some of it (for example, old glass and paper) is recycled, while the remainder is deposited at state dumps such as the one in the northern GDR at Schönberg.

THE POLICIES

If one were to judge the GDR by its laws, it would be an environmental leader. In 1970 a comprehensive environmental policy act, the *Landeskulturgesetz*,[40] was enacted; additional legislation has addressed specific issues in greater detail since then.[41] Numerous documents specify the legal rights and obligations of the various state bodies and establish the basis for emission limits of various pollutants as well as sanctions for violation of these limits. Environmental law is constantly evolving; leading legal experts in scientific institutions cooperate with central state agencies in drafting and amending legislation.

A basic principle underlying state policy is that the use, protection, and transformation of nature are inextricably linked. Environmental protection is not an end in itself, for natural resources are for human use. Therefore, the relationship between environmental protection and economic policy becomes pivotal. Most citizens of the GDR probably place a higher value on an improved material standard of living than they do on environmental protection, because compared to West Germany, the East Germans still believe that they are deprived of consumer goods and diet. The view expressed by the GDR economist Johann Köhler also probably predominates among economic practitioners: environmental protection measures "don't make us richer in material goods, but rather draw away working forces from the area of production of material goods."[42] In other words, if one must choose between environmental protection and economic growth, citizens and leaders alike would probably opt for the latter. Must one, however, choose between them?

Environmentally conscious economists and legal scholars in the GDR argue that economic growth and environmental protection are mutually reinforcing, not contradictory, goals; neglect of the environment will, in the long term, inhibit economic growth and thus hold back improvements in the standard of living. If the country's natural resource base is damaged, nature's "free services" (such as readily available pure water, the purification capacity and antierosion effects of forests, natural soil fertility, and natural decomposition of waste) must be replaced by expensive human-made processes (for example, purification facilities, chemical fertilizers and pest control, and extra health expenses).[43] The interdependence between economic growth and environmental protection is acknowledged also by political leaders, but with a different emphasis: an efficient economy assures adequate investment funds for environmental protection.[44] Nonetheless, the environmental perspective is becoming increasingly popular.

At the central level, the Ministry for Environmental Protection and Water Management established in 1971 determines the principal guidelines for environmental planning, especially water management; oversees implementation of laws; and makes proposals to the Council of Ministers regarding environmental questions. Productive enterprises and combines are required to include environmental protection in their short- and long-term plans. At the regional, district, city, and community levels, the local representative bodies and their executive councils also approve regional measures for environmental protection and, with the help of standing commissions and specialized technical organs, are responsible for coordinating and verifying the effectiveness of actions undertaken by enterprises in the region.[45] Other organs also play a role in overseeing implementation of state policy: the environmental inspectorate formed in 1985, the water inspectorate, the hygiene inspectorate, the fish health service, and the meteorological service, to name only some. Financial sanctions also encourage enterprises to make more careful use of natural resources.[46]

These complex legal and organizational structures should ensure an effective environmental policy. Policy often falters, however, at the stage of implementation. The fundamental problem is the relatively low priority of environmental protection, as compared to the goal of increased material production. Even if local organs are able to convince enterprises to include ambitious environmental projects in their preliminary economic plans, higher planning organs may water down or cut these projects altogether, for scarce supplies of labor, capital, and materials demand that choices be made. As a result, local projects that reinforce centrally established priorities (such as guaranteeing a supply of safe drinking water or improving the collection and utilization of waste) may well survive the complex planning process, while those that break new ground may not. The key question is this: who determines what is really practical and what is wishful thinking? Ultimately it is the Council of Ministers, and at this level productive priorities prevail most often. GDR specialists admit that effective environmental protection often depends on the personal initiative of the enterprise manager and on the active involvement of local state officials and citizens in pressing enterprises to respond to national policies. The law, so to speak, provides a lever, a countervailing force to tip the balance in favor of production over protection. If individual commitment is insufficient, environmental projects will most likely be sacrificed to other priorities.

The average citizen may complain by submitting petitions to the responsible authorities. These petitions are apparently quite common on environmental issues. By law, they must be answered by the appropriate state authority. They produce diverse reactions. In some cases they alert the authorities to unnoticed problems or draw attention to a situation that can be resolved within established priorities. At the local level, questions of location are frequent subjects of petitions. For example, a storage point for brown coal may be moved to a city's periphery in response to citizens' complaints about dust pollution in a residential area. A particular tree destined to be cut may be spared. In other cases, state policy may limit the ability of local officials to respond to petitions. An instance cited by a local GDR official may illustrate this difficulty: a heating plant to be constructed near a residential area initially posed no threat of pollution because it was to run on "clean" natural gas. When the countrywide shift to brown coal took place in the early 1980s, the plant was already under construction and could not be moved, even though now it was to be a coal-fired plant, producing high levels of pollution. Residents complained that the plant was too close to their homes. Little could be done, given the central commitment to brown coal. Ameliorative measures, such as a higher smokestack and dust barriers, were the only available options. In cases like this one, the official response to petitions is primarily an explanation of the necessity of the pollution. A petition sent to the state council of the GDR, the council of the Rostock region, and the council of the Grevemühlen district on 29 June 1986 met a similar fate. Distributed by organs of the Evangelical church in Berlin-Brandenburg and Mecklenburg, this petition called for an end to the importation of waste from the West and for full information, particularly in regard to the waste dump in Schönberg. The reply assured the petitioners that the Schönberg dump was safe and well regulated and not intended for toxic waste. Petitioners were invited to a meeting at the Ministry for Environmental Protection and Water Management for further explanation.[47]

Despite the constraints on effective citizen pressure, local state organs may be able to call some polluting enterprises to account. Because scientists and specialists on environmental matters may be members of the local standing commissions and technical organs responsible for environmental issues, these bodies sometimes command the expertise necessary to place effective pressure on enterprise officials. But some of these scientists are also undoubtedly part of a co-opted intelligentsia, that is, they benefit from the existing system in the form of higher salaries, opportunities for foreign travel, greater access to Western publications, and generous housing. Thus, they may air their criticisms in private or tone them down. This is not to say that there are no real environmentalists among the GDR's intellectuals. There are, and they apparently see themselves as a self-conscious pressure group sharing common goals.[48] Nonetheless, the East German leaders have generally been effective in keeping this type of environmentalism within limits where it endangers neither the system's stability nor its priorities.

THE CITIZEN

Popular attitudes toward environmental deterioration are difficult to assess, because survey research data are limited and GDR citizens may be reticent about expressing their views to foreign visitors. A West German scholar, Gerhard Würth, cites some interesting East German studies from the early to the mid-1970s that involved small opinion surveys of both adults and schoolchildren. Some 90 percent of the eight hundred persons questioned in Bitterfeld and Schwerin in 1971 believed air pollution to be damaging to health, and two-thirds of those in Bitterfeld wanted to move away, in large part because of the pollution. Of a group questioned in 1974, 30 percent were bothered by noise pollution.[49]

As the environment has become an ever more salient issue in West Germany, the East German citizens have also been exposed to more information from the Western media, even if only occasionally to reports on the GDR itself. The environmental question has also become an increasingly important theme in East German fiction.[50] Informal contacts with East German citizens and growing environmental activism within the GDR Evangelical church in the 1980s suggest a rising level of concern and at least a small dent in popular resignation. Even in the 1950s, societies for conservation of nature, in the narrow sense, served as outlets for interested individuals. Now numerous groups exist for those with a particular interest: hunting societies; fishing clubs; an organization of small gardeners, homesteaders, and breeders; and numerous branches of the *Kulturbund*, including ornithology and bird protection, nature and homeland, roses and orchids; and many others.[51]

With the passage of the *Landeskulturgesetz* in 1970, the regime has mobilized popular support for this law. The term "socialist *Landeskultur*" has a broader meaning than *Umweltschutz* (environmental protection) in the West. In addition to pollution control and nature conservation, it includes such diverse activities as urban planning, designation of protected cultural objects, and, in general, an improvement in living and working conditions. It is presented as a constructive process of environmental improvement, not simply the prevention or amelioration of environmental damage. The tone of public education and mobilization campaigns is positive, emphasizing the active involvement of the citizenry in improving the quality of life. Trade unions and professional organizations (of engineers, architects, and agricultural specialists, for example) also encourage their members to do their work with proper respect for ecological principles.[52] In neighborhoods and in the workplace, communal contracts with local state authorities establish citizens' duties for environmental improvement, such as responsibility for clearing ice in winter, help in maintaining recreational areas and parks, or participation in laying pipes for drinking water lines in rural areas. *"Mach mit"* ("participate") campaigns mobilize residents to beautify their cities and residential areas by taking better care of buildings, planting trees and shrubs, and cleaning up waste heaps. Local state organs also designate volunteer nature protection assistants to help in the planning, implementation, and enforcement of environmental policy at the local level under the leadership of volunteer specialists.[53]

This heavy reliance on volunteers reflects the scarce material and labor resources available within planning mandates. In 1980, mobilization of popular interest culminated in the formation of a new umbrella organization within the *Kulturbund* of the GDR—the Society for Nature and the Environment (SNE). With a membership of about sixty thousand in 1986, the SNE incorporated members from existing subgroups of the *Kulturbund* but brought a unifying theme and drew in additional members, including specialists (biologists, foresters, experts in water management), and in some cases even entire enterprises whose activities impinge heavily on the environment. For example, in the Dresden region, some twenty-five institutions and enterprises belong to the SNE. Organized on a regional basis, this society also has twelve specialized divisions including conservation, entomology, ornithology and protection of birds, hiking and tourism, and botany. Sections may make recommendations to state organs and otherwise organize a wide range of activities. For example, initiatives in Schwerin in 1980 involved taking care of parks, developing hiking trails, and organizing environmental exhibits. Specialized commissions carry out research projects on topics such as the condition of endangered species. A campaign initiated in the Dresden region mobilized 105,676 hours of volunteer work, worth an estimated 0.5 million marks. Subsequently, similar initiatives have been organized throughout the country under the same slogan: *"gepflegte Landschaft—gepflegte Umwelt"* ("a clean landscape means a clean environment").[54]

In part, organization of the SNE was intended to direct the energies of concerned citizens into channels that would reinforce rather than challenge official priorities. But one should not interpret such organizations exclusively as "transmission belts" of control from the top, because they also allow for the communication of citizens' interests to state officials and have helped to realize some citizen initiatives. While groups like the SNE clearly must operate within the parameters of state policy, control from above is probably much less rigid than one might expect.

Church Activities

Thus far we have focused on official environmental groups. Equally important, although much smaller in size, are semiautonomous activities organized within the East German Evangelical church. Since the mid-1970s, environmental issues have been of growing concern to many Christians, especially young people, and as churches became active, people having no connection to them were also drawn to these informal groups. The modus vivendi worked out with the state over the last several years has allowed the church a certain organizational autonomy in exchange for loyalty to the system. Therefore, the church has been able to serve as an umbrella for a wide variety of environmental—and peace—activities and discussions that embrace issues not aired in the official media. Young people find a freer atmosphere there, where they can more openly express their concerns. Church leaders also help them find ways to work within the system rather than seek emigration, fall into resignation, or engage in unacceptable forms of protest.

These small groups of environmental activists carry out diverse projects, sometimes described as "small steps." Although it may be impossible to bring fundamental change, these small steps taken together may make a difference, and—what is of equal importance to the participants—allow the individual to express personal commitment to an ecological lifestyle. Thus the ecological question takes on a spiritual and moral dimension. Activists question materialism and consumerism, implicitly suggesting alternatives to the regime's priorities. Young Christians discuss alternative lifestyles in which to seek fulfillment through "being" and not "having." Solidarity with the poorer nations of the Third World is presented as a corrective to the prevailing materialism. Church members in less polluted areas of the GDR have taken children from the highly industrialized centers in the south for holidays to allow them to recover from the excessive pollution.[55] Other small steps are meat-free days, car-free weekends, reduced use of household chemicals, organic gardening, fasts, and group tree planting.[56] The element of Christian asceticism in some of these actions is fundamentally at odds with the GDR's official values, and their spiritual foundation contradicts the materialism of Marxism-Leninism. Nonetheless, this spiritual emphasis may help make material shortages more palatable, and in this sense Christian environmentalism may not seem so threatening to the political authorities.

Like other citizens, church activists suffer from a lack of information about environmental conditions in their country. Thus church activists emphasize self-education and dissemination of available information through seminars, workshops, and exhibits. Personal contacts with scientists and individuals with special expertise are used to gain information. Hectographed church newsletters are sometimes distributed through the mail to interested individuals or passed hand to hand.[57] These newsletters, which represent a semiautonomous source of information subject to "self-censorship" rather than direct state censorship, are available in small numbers because of the difficulty in obtaining materials and printing facilities. Since January 1980, the Church Research Facility in Wittenberg (*Kirchliches Forschungsheim Wittenberg*) has produced a biannual "Letter for Orientation in the Conflict between Man and Nature," numbering twelve issues by October 1985. They include book reviews, information on local activities, reports on environmental damage, and essays on Christianity and the environment. Longer reports have emanated from work groups at Wittenberg and elsewhere in the GDR. Their topics have included agriculture and the environment, environmental protection in the household, forest damage, and alternative cooking recipes. Since 2 September 1986, an "environmental library" has been open (now three evenings per week) in a Berlin church, the Zionsgemeinde, and also serves as a location for informal discussion and regular lectures on environmental and other issues. A police raid on the library in November 1987, in which printing materials and some underground publications were confiscated and some activists were detained, may mark the beginning of intensified state control over such activities.[58]

Since the establishment of the official Society for Nature and the Environment in 1980, some church activists have also joined that organization. Their attempts to cooperate with state organs have met various fates. One tree planting organized by the church in 1981 failed when a state organization did not deliver the promised saplings, allegedly because of bad weather and insufficient size of the plantings. At the same time, the local Free German Youth organization *was* supplied with trees.[59] In other cases and especially in recent years, however, state organizations have seemed more willing to support these types of initiative.[60] They minimize the alienation of young people and, of course, trees are planted. But the state authorities look less favorably on other types of activities. For example, group bicycle tours of polluted areas may meet with official resistance: signs have been posted to prohibit bicycle traffic on the route and participants have been diverted. "Environmental worship services" in villages slated for razing for brown-coal strip-mining are also viewed unfavorably by the regime. Nonetheless in 1985 a representative of the regional council reportedly was available to answer questions after a service in Potzschau, south of Leipzig.[61] In Erfurt, the Society for Nature and the Environment now allows church activists to serve as volunteer water protection commissioners,[62] in an apparent attempt to harness the constructive energies of these young environmentalists instead of suppressing them. As long as these activities remain clearly under the umbrella of the church, further official goals, and do not "take to the streets" giving the appearance of a demonstration, they seem to operate with relatively little interference.

The Chernobyl accident inspired new initiatives inside the church. On 5 June 1986, 140 citizens addressed a petition to the council of ministers containing a fundamental criticism of nuclear power. The petitioners demanded better information on the dangers of nuclear power, clear data on post-Chernobyl radiation levels, an end to construction of new plants, a reorientation of energy policy so as to end reliance on nuclear power by 1990, increased research on alternative sources of energy, and improved energy conservation.[63] Later in June, a petition supported by thousands was delivered to the legislature, this time demanding a referendum on the further development of nuclear power.[64] Church activists have also sought to gather and disseminate available information among themselves. These actions represent a marked contrast to the pre-Chernobyl era, when antinuclear sentiment was rarely aired.

Although church-related activism involves limited numbers of individuals, the moral commitment of those who do participate gives it a greater social weight than numbers suggest. For unlike the larger official organizations, these small semiindependent groups challenge the information monopoly of the state and stimulate creative initiatives; they capture both the energies and the imagination of talented young Germans.

CONCLUSIONS

The impact of pollution often goes undetected until the damage is done; furthermore, even scientists with sophisticated research facilities disagree over its long-term effects on health and well-being. Therefore, the East German citizens, hindered by poor information and surrounded by reassuring messages from the authorities, may find environmental problems less pressing than many other problems. Housing shortages, the difficulty of finding the thousand little things needed for everyday life, worse air and greater noise at work than at home, and limited possibilities for foreign travel may all prove more frustrating day to day than the invisible effects of nitrates and heavy metals in food and water, endangerment of species, or polluted groundwater. To be sure, some environmental damage *is* becoming increasingly obvious: malodorous air and dirty laundry, rivers and lakes that are closed to swimming, damaged forests, and dead fish. But many citizens probably see these as largely unavoidable costs of progress against which one can do little, if anything.

Both the GDR and Western media make it clear that socialism has no monopoly on environmental damage: localized environmental catastrophes are now regular affairs in capitalist and Third World countries. Even if the East German citizens know that domestic problems go unreported, the news blackout keeps this danger on an abstract level. Unlike in Western countries, in the GDR the citizens do not learn about a specific toxic waste hazard next door or about measured contamination in food and water. The dangers are theoretical and abstract, while the shortages of fresh fruits and vegetables are real and visible. Perhaps environmentalism is a luxury of the rich.

The costs of activism may also seem high, perhaps higher than they really are. Work in official organizations is praised and not censured, and one can even participate in some church activities without major consequences. Yet if one oversteps the bounds by trying to organize too independently or learning too much, there are penalties to career and education, in some cases even imprisonment.

The environmental awakening is arriving slowly in the GDR, but it is arriving. The question in the GDR is the same as in the West. Will citizen and regime realize and respond to the dangers before they are irreversible?

NOTES

This chapter was previously published in 1989 in *The Quality of Life in the German Democratic Republic: Changes and Developments in a State Socialist Society*, edited by Marilyn Rueschemeyer and Christiane Lemke. It is reprinted, with minor changes, by permission of the publisher, M. E. Sharpe (Armonk, New York; London, England). The author wishes to thank McGill University, the International Research and Exchanges Board, and the Canadian Social Sciences and Humanities Research Council for support in carrying out this research.

1. Based on information from the office of the city senator for urban development and environmental protection, West Berlin. For regulations governing smog in West Berlin see "Verordnung zur Verminderung schädlicher Umwelteinwirkungen bei austauscharmen Wetterlagen (Smog-Verordnung). Vom 25. Oktober 1985," *Gesetz und Verordnungsblatt für Berlin* 31 (12 November 1985): 2282–84.

2. Radio interview with Dr. Helmut Breitenkamp of the Referat Luftreinhaltung of the Berlin Senate, transmitted on RIAS I (West Berlin), 8 February 1987, 6:35– 7:15 a.m. Interview conducted by Martin Irion.

3. Peter Wensierski, *Von oben nach unten wächst gar nichts: Umweltzerstörung und Protest in der DDR* (Frankfurt am Main: Fischer Taschenbuch Verlag, 1986), p. 23 (hereafter cited as Wensierski).

4. See, e.g., *epd* (evangelisches pressedienst), Landesdienst Berlin, no. 113, 18 June 1984; and "Mehr kommunizieren als konsumieren," *Die Kirche*, 14 July 1985.

5. *Statistisches Jahrbuch der Deutschen Demokratischen Republik 1986* (Berlin: Staatsverlag der DDR, 1986), pp. 155, 257.

6. "Luftverunreinigung in der DDR: Die Emission von Schwefeldioxid und Stickoxiden," *DIW Wochenbericht* 52: 30/85 (25 July 1985): 342.

7. Interview with Cord Schwartau of the Deutsches Institut für Wirtschaftsforschung (DIW) in West Berlin, 11 March 1987. See also "Emissionen von SO_2 aus Braunkohlekraftwerken in der DDR," *DIW Wochenbericht* 54: 11/87 (12 March 1987): 154–57.

8. *DIW Wochenbericht*, no. 30/85, p. 337.

9. *DIW Wochenbericht*, no. 11/87, p. 155.

10. Karlwilhelm Horn, "Vorwort," *Wissenschaftliche Zeitschrift der Humboldt Universität Berlin*, Mathematisch-naturwissenschaftliche Reihe, no. 5 (1970).

11. Marlies Menge, "Für Filter fehlen die Devisen," *Die Zeit* (Hamburg) (Canada edition), 25 March 1983, p. 21; and Gerhard Würth, *Umweltschmutz und Umweltzerstörung in der DDR* (Frankfurt am Main: Peter Lang, 1985), pp. 68–72 (hereafter cited as Würth).

12. G. W. Dominok and R. Schweissinger, "Zur Epidemiologie der 10 häufigsten Malignome in Bezirk Cottbus," *Das deutsche Gesundheitswesen* 39: 30 (1984): 1770–71.

13. "Experte hält 90 Prozent des Waldes der DDR für geschädigt," *Tagesspiegel*, 17 February 1987.

14. See *DIW-Wochenbericht*, no. 11/87, pp. 155–57.

15. Ibid., p. 157; and Michael Schmitz, "Der ungeteilte Dreck," *Die Zeit*, 6 March 1987, p. 42.

16. Speech by Hans Reichelt, "Der Schutz der Umwelt erfordert Entspannung im Geist von Helsinki," *Neues Deutschland*, 9 July 1985, p. 3.

17. Würth, p. 41.

18. Manfred Melzer, "Wasserwirtschaft und Umweltschutz in der DDR," ed. Maria Haendcke-Hoppe and Konrad Melzer, *Umweltschutz in beiden Teilen Deutschlands* (West Berlin: Duncker und Humblot, 1985), p. 79 (hereafter cited as Melzer); and Würth, pp. 248–51 (hereafter cited as Haendcke-Hoppe).

19. Erich Honecker, "Bericht des Zentralkomitees der Sozialistischen Einheitspartei Deutschlands an der XI. Parteitag der SED," 17–21 April 1986 (Berlin: Dietz Verlag, 1985), p. 31 (hereafter cited as Honecker); see also "Die Kernenergie der RGW-Länder," *DIW Wochenbericht* 53: 25/86 (19 June 1986): 309–10.

20. See Wensierski, p. 92.

21. See Joan DeBardeleben, "Esoteric Policy Debate: Nuclear Safety Issues in the Soviet Union and German Democratic Republic," *British Journal of Political Science* 15 (1985): 227–53.

22. For a discussion of the evidence see Wolfgang Mehringer, "Reactorsicherheit in

der DDR," *IGW-report über Wissenschaft und Technologie in der DDR und anderen RGW-Ländern*, no. 1, Institut für Gesellschaft und Wissenschaft, Universität Erlangen-Nürnberg, 1987, pp. 34–52.

23. See *Neues Deutschland*, 29 April 1986, p. 5; 2 May 1986, p. 2; 3–4 May 1986, p. 1; 5 May 1986, p. 1; and 6 May 1986, p. 5.

24. *Neues Deutschland*, 3–4 May 1986, p. 1.

25. *Der Tagesspiegel* (West Berlin), 13 February 1987, p. 2.

26. Ibid.

27. Honecker, p. 31.

28. See Andreas Kurjo, "Landwirtschaft und Umwelt in der DDR: Ökologische, rechtliche und institutionelle Aspekte der sozialistischen Agrarpolitik" (hereafter cited as Kurjo), *Umweltprobleme und Umweltbewusstsein in der DDR*, Deutschland Archiv (Cologne: Verlag Wissenschaft und Politik, 1985), pp. 60–63 (hereafter cited as *Umweltprobleme*).

29. Karl Hohmann, "Die Industrialisierung der Landwirtschaft und ihre Auswirkung auf die Umwelt in der DDR," in Haendcke-Hoppe, pp. 61–63, 65 (hereafter cited as Hohmann).

30. Based on material provided to the author by Peter Wensierski from his forthcoming book (coauthored with Wolfgang Büscher), *Ökologische Probleme und Kritik an der Industriegesellschaft in der DDR heute*.

31. See, e.g., *Berliner Morgenpost*, 1 March 1987, p. 1; and Hohmann, pp. 56–57.

32. Kurjo, pp. 66–67.

33. Hohmann, pp. 45–46.

34. "Natur und Landschaftschutz in der DDR," *Presse-Informationen*, Presseamt beim Vorsitzenden der Ministerrats der DDR, no. 146 (17 December 1986), p. 6.

35. Würth, pp. 211–14. As Würth points out, little concrete data are available for the past twenty years. See also Melzer, pp. 73–80, 84–85.

36. Eberhard Garbe and Dieter Graichen, *Sekundärrohstoffe: Begriffe, Fakten, Perspektiven* (Berlin: Staatsverlag der DDR, 1986), pp. 202–4.

37. Ibid., p. 77.

38. Ibid., pp. 74, 90, 11, 147; Hans Reichelt, "Die natürliche Umwelt rationell nutzen, gestalten und schützen," *Einheit* 39 (November 1984): 1013 (hereafter cited as Reichelt).

39. *Landeskulturrecht*, ed. Ellenor Oehler (Berlin: Staatverlag der DDR, 1986), pp. 202–4 (hereafter cited as *Landeskulturrecht*).

40. "Gesetz über die planmässige Gestaltung der sozialistischen Landeskultur in der DDR," *Gesetzblatt der DDR*, pt. I, no. 12 (1970), pp. 67ff.

41. For an overview of environmental law, see *Landeskulturrecht*, 1986.

42. Johann Köhler, "Zu den Problematik der produktiven und der unproduktiven Arbeit sowie der Dienstleistungen," *Wirtschaftswissenschaft* 22 (June 1974): 886.

43. See, e.g., Ellenor Oehler et al., *Landeskulturrecht: Lexicon*, s.v. "Gratisdienste der Natur" (Berlin: Staatsverlag der DDR, 1983), p. 67 (hereafter cited as *Landeskulturrecht: Lexicon*); and Horst Paucke and Günter Streibel, "Gratisdienste der Natur," *Wirtschaftwissenschaft* 31 (September 1983): 1317–32.

44. See, e.g., "Internationale Konferenz zum Umweltschutz in Helsinki," *Neues Deutschland*, 9 July 1985, p. 1; and Reichelt, p. 1012. For a detailed discussion of the relationship between economics and ecology in the GDR see Joan DeBardeleben, *The Environment and Marxism-Leninism: The Soviet and East German Experience* (Boulder, Colo.: Westview Press, 1985), chaps. 5–7 (hereafter cited as DeBardeleben).

45. *Landeskulturrecht*, pp. 51–52.

46. For further discussion of these measures and citations of relevant legislation, see DeBardeleben, pp. 159, 246; and *Die staatliche Leitung der Bodennutzung: Rechtsfragen*, ed. Ellenor Oehler (Berlin: Staatsverlag der DDR, 1985), pp. 95–96.

47. The response to the petition is summarized in a newsletter, "Die Umweltbibliothek," by an environmental group in the Zionsgemeinde, Berlin, October 1986.

48. This impression is based on interviews with scientists in the GDR.

49. Würth, pp. 63–67.
50. For an overview see Hubertus Knabe, "Der Mensch mördert sich selbst," *Deutschland Archiv* 16 (1983): 954–73. For a recent example see Christa Wolf, *Störfall* (Berlin, Weimar: Aufbau Verlag, 1987), also published in the FRG (Darmstadt: Luchterhand, 1987).
51. See *Handbuch gesellschaftlicher Organisationen der DDR*, ed. Richard Mand et al., pp. 43–44, 97–98, 107–10, 160–64, 188.
52. Ibid., pp. 23, 34–36, 98–101.
53. *Landeskulturrecht: Lexicon*, 1983, pp. 124–25.
54. Peter Wensierski, "Die Gesellschaft für Natur und Umwelt," *Umweltprobleme*, pp. 151–68; Manfred Fiedler, "Initiativen für Umwelt und Natur," *Einheit* 39 (November 1984): 1024–27; and *Mitteilungen für die Staatsorgane in Bezirk Dresden*, no. 3 (1986), p. 13.
55. "Aktion 'Saubere Luft' für Ferienkinder," *Frieden und Freiheit*, June 1985; and Wensierski, p. 44. On the spiritual dimension, see DeBardeleben, pp. 87–91, 97–98, 189.
56. See Wensierski, pp. 161–94; and Detlef Urban, "Die Umweltarbeit der Kirchen," *Kirchen und Gesellschaft in beiden deutschen Staaten*, ed. Gisela Helwig and Detlef Urban (Cologne: Verlag Wissenschaft und Politik, 1987), pp. 131–36 (hereafter cited as Urban).
57. DeBardeleben, pp. 79, 87–91; and Detlef Urban, "Kirchen treten an die Öffentlichkeit," *Die evangelischen Kirchen in der DDR*, ed. Reinhard Henkys (Munich: Chr. Kaiser Verlag, 1982), pp. 341–49.
58. *New York Times*, 29 November 1987, p. A24.
59. "Aus einem Interview in einer Jungen Gemeinde," in Urban, p. 132.
60. "Bericht eines Mitgliedes des Arbeitskreises Umweltschutz-Eisenach," in Urban, p. 133; and "SED würdigte Engagement der Kirchen in Ökologiefragen," *Frankfurter Rundschau*, 11 September 1984.
61. "Mehr kommunizieren als konsumieren," *Die Kirche*, 14 July 1985.
62. Gesine Schmidt, "Bericht über praktische Arbeit von Ökogruppen," in Wensierski, p. 166.
63. Portions of the appeal were published in West Berlin's *Tageszeitung*, 23 June 1986.
64. See "DDR-Bewohner fordern ein Umdenken in der Energie- und Informationspolitik," *Frankfurter Allgemeine Zeitung*, 24 June 1986.

10

RECENT HUNGARIAN APPROACHES TO AGRICULTURAL POLLUTION

Zoltán Király

Hungarian experts in environmental and land use policy and agricultural economics have come to the conclusion that temporary advantages gained from the intensive exploitation of natural resources may also entail major disadvantages. In many cases the uncontrolled use of technology has turned out to be harmful to the natural environment. For instance, current agricultural practices have brought not only high yields but also the problem of pollution. High growth rates in agricultural productivity in Hungary, as elsewhere, are linked with some undesirable consequences for the natural environment and public health.

THE CURRENT SITUATION AND PROBLEMS THAT NEED TO BE SOLVED

Plant Protection

The number of pesticides, including the number of active ingredients they contain, has increased during the past few decades. The toxicity of the pesticides being used today in Hungary has declined in comparison to those that were applied forty years ago, however. According even to pessimistic extrapolations, the toxicity of pesticides will continue to diminish.

In the past twenty years many mistakes have been made in the manufacture and application of pesticides against plant diseases, weeds, and insect pests. Especially important are the nontarget effects of DDT and 2,4,5-T, which have been revealed to be harmful even to humans. Recent tragic accidents such as the release of methyliso-cyanate (a toxic intermediate used in the manufacturing of pesticides) in Bhopal, India, in 1984 have called attention to the importance of health and environmental precautions that should accompany plant protection. A recent Canadian study found a strong correlation between the incidence of Parkinson's disease and the level of herbicide use. Herbicides containing paraquat or cyperquat and some structurally similar defoliating agents were implicated as a cause of the disease.[1] In 1985, the Pesticide Action Network International launched a "Dirty Dozen" campaign. The campaign targets twelve particularly hazardous pesticides known to cause poisonings, food contamination, and severe environmental problems. The goal of this campaign is to halt the use of these pesticides and to replace them with environmentally safe control measures. The "Dirty Dozen" are DDT, EDB, 2,4,5-T, DBCP, chlordane/heptachlor, toxaphene, paraquat, parathion, BHC/lindane, pentachlorophenol, the "drins" (aldrin, dieldrin, and endrin), and chlordimeform.

Many of the "Dirty Dozen" have never been introduced in Hungary. DDT, BHC, and the "drins" were banned as early as 1968. Hungary was the first country to stop using some of these pesticides and replace them with others. Mercuric compounds and 2,4,5-T also have not been in use for a long time. Paraquat-containing pesticides are the only ones of the "Dirty Dozen" still being used. Gramoxone (a weed killer containing paraquat) and the defoliating agent Reglone are regularly applied in corn, potato, and sunflower fields. It would be very difficult at present to replace Gramoxone or Reglone with other compounds.

The use of toxic pesticides is permitted by the Ministry of Agriculture and Food on collective and large state farms only to experts with licenses to handle poisonous compounds. On these large farms, "plant protection engineers" who hold college or university diplomas are responsible for plant protection management. (Students who complete five semesters at the Agricultural University in plant pathology, entomology, weed research, pesticide chemistry, practical plant protection, and economics of control measures receive the diploma of "plant protection engineer.") These plant protection engineers are usually well acquainted with schedules of control measures, new pesticides, toxicological aspects of crop protection, and the economics of control. In the past six years not a single case of acute pesticide poi-

Table 10.1
PESTICIDE DOSES UNHARMFUL TO SOIL BACTERIA AND DOSES REGULARLY ADMINISTERED

Active ingredient	Unharmful dose (kilograms per hectare)	Regular dose (kilograms per hectare)	Safety factor[a]
Lindane	1760	1.5	1173
Ethylparathion	100	0.5	200
DNOC	175	2.0	87
2,4-D	1000	2.0	500
Atrazine	200	1.5	133

SOURCE: F. Hargitai, lecture presented at the Hungarian Academy of Sciences, Agricultural Section, 7 May 1986.
[a]Unharmful dose divided by regular dose.

soning has occurred at a state or collective farm. But the situation on smaller farms as well as hobby farms and gardens is not equally sat isfactory, for hobby farmers have a limited knowledge of the proper use of pesticides. Because chemicals are usually not available in small quantities, small farmers and hobby gardeners often store large quantities of poisonous pesticides at home, where they can be dangerous, particularly for children.

Evidence suggests that so far the use of pesticides and fertilizers has not caused major damage in Hungary. Groundwaters, soil waters, drinking waters, lakes, and rivers are not at present polluted with pesticides or fertilizers. This conclusion was included in a 1986 report titled *The Prevention of Non-Point Source Pollution* from a project supported, organized, and carried out in Hungary with the aid of the Farm and Agricultural Organization—United Nations Development Program.

Soil bacteria, with the exception of *Rhizobium* species, have not been eliminated through the use of pesticides (table 10.1). The safety factor for most pesticides is rather high, which means that in most cases even doses several hundred times greater than those normally used would not be harmful, given the number of bacteria in the soil.

A ten-year survey monitoring insect communities in apple and corn stands under chemical pesticide treatment was recently completed at the Plant Protection Institute in Budapest. It showed that even in intensively treated agricultural areas the numbers of both harmful and beneficial insects are reduced only temporarily, and they rapidly reestablish themselves from the nontreated areas. In the ten-year period, no species became extinct.[2]

Despite these findings, it must be stressed that it is difficult to detect and evaluate the nontarget effects of pesticide use. In many instances only incidental experiences called our attention to them. Recently, for instance, "pesticide rain" containing BHC was reported from Hokkaido in Japan because of the heavy use of BHC in China and Korea.

In Hungary today, 11 kilograms of pesticides are applied annually per hectare of arable land. The active ingredients constitute roughly half of that quantity, or 5.5 kilograms per hectare. In vineyards and orchards, however, 10–15–18 pesticides have been regularly applied, so that the pesticide load per hectare may be around 60 kilograms per hectare instead of the average of 11 kilograms per hectare. The proportion of pesticides being used in Hungary is as follows: herbicides 60 percent, fungicides 30 percent, insecticides 10 percent. The most common chemical compounds belong to the herbicide group.

Neither the public nor the government is satisfied with the conventional measures of chemical control. The public, however, tends to exaggerate the promise of new biological and other nonchemical control measures. Several slogans reflecting concern with the quality of the environment have emerged in the past twenty years, for example: "Environmentally safe, environmentally sound, integrated, nonchemical, alternate, selective, and biological plant protection." Although we cannot expect a general breakthrough in substituting herbicides with nonchemical control measures, it is important to reduce chemical pesticides in as many fields as possible.

Fertilization and Acidification of Soils

Almost 50 percent of Hungarian soils have low pH values, while soil acidification is very intense in 30 percent of the soils. Furthermore, more than 50 percent of Hungarian soils are deficient in carbonates,[3] and the nitrogen-producing capacity is insufficient in 70 percent of the soils. This means that nitrogen, mostly in the form of fertilizers, is badly needed. The phosphorus-producing capacity is very good in 30 percent, good in 34 percent, and weak or medium in 36 percent of soils. The potassium-producing capacity is weak in 50 percent of soils; these soils too need fertilizer. Neutral soils yield 20 to 25 percent more crops than acidic ones, as was indicated in experiments with wheat, corn, and sunflower.[4] There is also a cause-and-effect relationship between the nitrogen-phosphorus-potassium (NPK) capacity of soils and their yielding capacity.

About ten times more fertilizer is used in Hungary today than was used twenty-five years ago, or about 280 kilograms of NPK active ingredients per hectare. This figure approaches West European levels and is much higher than the average 90 kilograms per hectare in both the United States and the USSR. The tenfold increase in fertilizer use paralleled higher crop yields. The increased agricultural production is one of the main factors underlying the improved living standard in Hungary, making the maintenance of high-input intensive agriculture a national priority. High doses of fertilizer have, however, been implicated in several problems caused by environmental pollution and soil acidification. Indeed, if one compares the average soil pH in Hungary in 1980 (6.47) to the 1985 average (6.10), the acidification trend is obvious. It has continued since 1985. The calcium carbonate content, on the other hand, decreased in the investigated soils (table 10.2), contributing to the acidification and justifying the need for lime treatment of soils.

Table 10.2
CALCIUM CARBONATE CONTENTS OF HUNGARIAN SOILS, 1978–81 AND 1982–85

Percentage of Hungarian Soils	Percentage of Calcium Carbonate			
	0.5	0.5–1.0	1.0–5.0	5–15
1978–81	49.5	6.7	25.8	18
1982–85	55.9	5.3	20.8	18

SOURCE: "Proposal for the Ministry of Agriculture and Food," Budapest, 1986.

The consequences of intensive fertilization are manifold. Though the source of high nitrate levels in drinking water and crop plants in some areas is disputed (candidates include leakage of nitrogen fertilizers, wastes from livestock, and the disposal of human wastes), it seems likely that they are the cause of hemoglobinemia in infants.

Cadmium pollution is another risk. It is mainly superphosphate that contributes to cadmium pollution because it contains 0.5 grams of cadmium per 100 kilograms of fertilizer. Reducing the cadmium content of the superphosphate fertilizer and diminishing its acidity through ammonification are urgent tasks for the Hungarian chemical industry.

The increased use of fertilizer may also be associated with unnecessary solubilization of several metal elements (nickel, lead, chromium, cadmium, and zinc), which can reach poisonous levels in food crops. Poisonous levels occur because the acidification of soils increases with the high use of fertilizers or with acid rain. Another possible impact of acidification is that high levels of some elements in crop plants may increase their susceptibility to pathogens. Some elements, such as zinc, may also stimulate the virulence of infectious microorganisms. It is well known among plant pathologists that plants grown on low-pH soils are attacked heavily by *Fusarium* fungi. All these pollutants are dangerous primarily in sandy soils that contain low levels of calcium carbonate. Elevated nitrate concentrations occur in groundwaters and drinking waters mainly in sandy soils.

The acidification of the environment, particularly of soil, has raised several questions and stimulated intense debate over causes and effects. Two factors seem to be important: acidic depositions from the air and increased use of fertilizer. But the exact role of these factors in causing acidification is unknown. Acid rain may be caused by the sulfur dioxide and nitrogen oxide contents in the air. Inland deposition of sulfur dioxide in the form of acid rain and dry deposition seem to be high. To neutralize the sulfur dioxide content in acid rain in one hectare of arable land would require 35 to 280 kilograms of calcium carbonate, depending on average precipitation. Various nitrogen oxides also contribute significantly to the acidification of the environment. The exact role of fertilizer use in acidification is difficult to determine. In Hungary, a connection has been made to nitrate- and phosphorus-containing fertilizers, but satisfactory evidence is not available. Nonetheless, one cannot rule out the contribution of fertilizers to the acidification of soils and other environmental elements.

NEW APPROACHES TO THE PROBLEM

Plant Protection

The only realistic approach to pollution problems, at least in Hungary, is to balance chemical and nonchemical methods to manage plant diseases, pests, and weeds.[5] Food production must continue to grow through "high input" agriculture. Although no breakthrough may be expected in nonchemical plant protection in the next twenty years or so, a slow decline can be anticipated in the use of chemical control measures and increased efficiency in environmental protection.

Weed Control

Crop production in Hungary still relies on chemical weed killers. Most likely, chemical weed control measures will not be replaced by biological or other nonchemical methods in the near future.

At least three developments deserve mention:

1. According to the results of experiments conducted in the Plant Protection Institute in Budapest, phospholipid-containing adjuvants (which are derived from soybean or sunflower seeds) in a mixture with herbicides can reduce the effective concentration of postemergent herbicides. The adjuvants increase the penetration of the weed-killing compounds into the different organs of weeds, thereby reducing the concentrations of poisonous herbicides.
2. Glyphosate herbicides can be mixed with fertilizer containing ammonium nitrate and sprayed on plants in the field.[6]
3. A revolutionary method developed by the Du Pont Company involves the release of a new sulfonylurea-type herbicide that, in extremely low concentrations of a few grams per hectare, can kill weeds in cereal crops. The extremely low quantity of this herbicide means that it produces virtually no environmental pollution, and the toxicity of the new herbicide for animals and humans is no higher than that of regular herbicides; thus the safety of this new compound appears guaranteed. The Organic Chemistry and Weed Research departments of the Hungarian Plant Protection Institute have a promising program with their own low-concentration herbicide for wheat.

These approaches are the only realistic methods, at least in Hungary, that can be applied in crop production to reduce herbicide pollution in agriculture and horticulture.

Plant Diseases

The control measures for controlling crop diseases caused by infectious microorganisms include several biological (nonchemical) methods and programs of integrated management:

1. In the 1930s scientists already realized that a few soil fungi (for example, *Trichoderma*) produce antibiotics that inhibit pathogenic soil fungi that cause damping-off disease and root diseases.[7] Several species of *Trichoderma* are antagonistic to fungi such as *Pythium*, *Rhizoctonia*, and *Fusarium*. Researchers have shown that propagules of *Trichoderma* introduced into the soil produce antibiotic substances, and these applied as seed coat can control some major plant diseases. In recent years there has been a

renaissance of *Trichoderma* research because of pressure from environmentalists. The Hungarian endeavors are also promising.[8]

2. Infectious wilt diseases of plants (including fusarial infections) are difficult to control by either chemical or nonchemical means. An integrated method, however, worked out in the Hungarian Plant Protection Institute, has proved very successful.[9] Research indicates that juvenile plant tissues are resistant to symptoms caused by nectrotrophic pathogens.[10] It is possible to increase plant juvenility in several ways. The most suitable method is to supply abundant nutrients, especially nitrogen, in the soil or in the nutrient solution. Particularly nitrogen in the form of nitrates causes juvenility in several crop plants. Leaves become green, the chlorophyll content increases, and the cytokinin hormone level rises in the tissues. The composition of membranes also changes because the phospholipid-to-sterol ratio increases.[11] All in all, the high nitrate-nitrogen supply eliminates the need for fungicides in controlling fusarial wilt diseases.

3. It has been shown recently that fungicides in sublethal dosages can substantially affect the development of diseases. Tests of triadimefon against powdery mildew of wheat determined that disease efficiency and sporulation capacity were reduced greatly by amounts less than one one-hundredth of the recommended dosages.[12] Fungicides used in this way may make it possible to manage some plant diseases at lower cost and with a reduced risk of nontarget effects.

Insect Pests

In controlling insect pests several biological and selective control measures are available to reduce the use of insecticides:

1. The most important prerequisite is an effective forecast system. Successful forecasting of the mass occurrence of insects helps to reach two goals: it reduces the number of chemical sprayings and treatments and at the same time preserves the natural enemies of insect pests, minimizing environmental pollution. Furthermore, if the timing of chemical control is good, populations of insect parasites and predators will not be damaged, and thereby the natural balance between insects and their enemies will be preserved. Forecasting the natural enemies of insects also diminishes the nontarget effects of insecticides. During the past

few years a joint project of the Plant Protection Institute and the Plant Protection and Agrochemical Service of the Hungarian Ministry of Agriculture and Food established a system for fore-casting insect infestations in apple orchards so as to preserve the natural enemies of apple insects. This method involves precise timing of chemical sprayings or even omitting them altogether during the peak presence of these enemies.

2. By applying color and other traps in orchards one can effectively reduce the insect population, such as the cherry fruit fly, without chemicals. Hungarian entomologists have concluded that it is possible to produce cherries without any chemical controls if color traps are used. There is, however, a regular 5 percent infestation of cherry fruit flies in the fruits, a figure acceptable to environmentalists but not to consumers.

3. Traps are sometimes combined with attractants and pheromones that modify insect behavior. In Hungary, traps are used mainly to register changes in insect populations to provide a basis for a reliable forecast system.

The selective chemical control of insects comprises insect hormone analogues, pheromones, and specific attractants or repellents that are not harmful to other living organisms, including people. These very specific compounds interfere only with specific insect functions or with the synthesis of compounds, such as chitin or juvenile hormones, which are formed only or mainly in the insect pests. These very selective insecticides are environmentally safe and have no harmful impact on human health. Experts are optimistic about the future of selective control measures despite the fact that at present they are only in an experimental stage.[13] Pheromones are already being used for monitoring and trapping boll weevil in cotton farms where insecticides are applied only when needed to hold down the population, as well as in forests where it is not practical to use pesticides but where trapping, attractants and repellents, and biological control measures look promising.

Genuine biological control of insects has a long history. Hyperparasites (*Trichogramma*), predators, and insect disease-causing bacteria such as *Bacillus thuringiensis* or fungi (*Beauveria bassiana*) may be used. In Hungary a few pheromones, *Bacillus thuringiensis*, hyperparasites, and predators in the greenhouse are used for plant protection. *Bacillus thuringiensis* is effective against insect pests on forest trees (tussock moth) and against noctuids and arctiid moths, which damage several plant species. The parasitic microwasps (*Trichogramma*) lay their eggs inside the eggs of several insect pests, thereby limiting the numbers of harmful insects. One of the most promising fields of application of *Trichogramma* would be the biological control of the European corn borer. Unfortunately, several technical requirements for the mass production and distribution of parasitic wasps cannot be met at present.[14]

Breeding for Resistance and Producing Disease-Resistant Cultivars

Breeding and producing disease- and pest-resistant crop cultivars seems to be one of the most effective nonchemical control measures. So far, moderate but important successes have been achieved in this field. Although not a genuine control measure, this is the most effective "biological control" of diseases and pests of crop plants. In the near future several crop plant cultivars that are also resistant to or tolerant of herbicides will probably be released. This release will be made possible by the application of new biotechnological methods. Some breeding methods such as selection, sexual hybridization, and the induction of mutations have been used in the past. Recently, the new somatic genetics, the recombinant DNA technique, and several biotechnological procedures have made it possible to create and select new plant species.[15]

Vegetative selection methods or in vitro selection techniques permit the selection of protoplasts, single cells, or tissues against pathogenic toxins and herbicides.[16] The main problem lies in regenerating the resistant protoplasts, cells, or tissues into whole plants. Current possibilities are rather limited. Another method of breeding for resistance to disease is the somatic hybridization of protoplasts or cells, which transmits genes for resistance from one species to another. This method seems particularly useful when the two species to be crossed cannot be hybridized by sexual crossings. In applying this method, the main problem again lies in regenerating the whole plant.

The recombinant DNA technique seeks to introduce new genes into crop plants by applying several vectors such as *Agrobacterium tumefaciens*, *A. rhizogenes*, or some DNA plant viruses.

It was found recently that direct genes could be transferred to plant protoplasts, eliminating the need for introducing the desired foreign gene into *Agrobacterium* or other vectors before inserting it into the plant. Protoplasts take up DNA, which is rarely able to penetrate the cell wall, and only with very low efficiency. Using a new method, plant protoplasts are treated with a high-voltage electric pulse (electroporation) that alters their membranes, allowing a direct introduction of the foreign gene into the plant protoplast. This procedure has been combined with heat shock and other procedures to achieve high transformation efficiencies.[17] The transformants need to be regenerated into whole plants.

These new biotechnological procedures offer great promise for breeding disease-, insect-, and herbicide-resistant plants, thereby reducing or altogether eliminating the need for chemical control measures.

The following features of present or possible policy for plant protection in Hungary should be noted:

1. For each poisonous pesticide the exact withdrawal time must be calculated. This is the time during which the potential pesticide content of food plant tissues is decomposed.
2. Safety circles have been marked around the larger lakes within which the use and storage of pesticides is not permitted.
3. The list of pesticides available to small farmers and hobby gardeners will be revised and some of the safer compounds will be made available in small quantities. Hobby farmers cause the greatest pollution in proportion to application of pesticides.
4. Plant protection drugstores have recently started operating to advise private farmers and hobby gardeners on the application of pesticides. These new stores are located in the vicinity of plant clinics already existing within the framework of local branches of the state plant protection laboratories.
5. The number of applications of pesticides in the large state and collective farms could be reduced if forecast systems were made more effective. Additional research is needed.

Fertilization

The following measures must be taken, or are now being taken, to reduce the damage caused by the application or improper use of fertilizers:

1. A liming program to reduce the unfavorable effects of soil acidification. This is partly carried out as a melioration liming program and partly for the continuous maintenance of neutral soil pH. About 50 percent of the costs are already covered by the state.
2. Introduction of the tram-line system into the cultivation of crop plants so that small dosages can be applied. This system will probably reduce to a tolerable level the leakage of fertilizers,

particularly nitrates (and other nutrients as well as pesticides), from sandy soil.

3. Changes in manufacturing fertilizers to diminish the acidification of soil. At present 40 percent of fertilizers containing phosphorus are already of the monoammonphosphate compound type instead of the acidic superphosphate.

4. Research on the possible protection of plants from the action of pollutants such as sulfur dioxide and pesticides as well as the action of acidic soil and acid rain. Several plant hormones such as cytokinin and abscisic acid are good candidates as stress protectants.[18]

Despite progress in many areas, it must be stressed that increased agricultural productivity in Hungary will require the use of a wide range of chemical and biological inputs (fertilizers, pesticides, and breeding of resistant and high-yielding plant cultivars) in the future as it has in the past. This "high input" agriculture will remain the principal means of increasing food production for a few decades. The challenge in our time is to harmonize this concept and practice with the protection of the agricultural environment from pollution.

NOTES

This chapter is based on a paper presented at the conference "Environmental Problems and Policies in Eastern Europe," sponsored by the East European Program of the Woodrow Wilson International Center for Scholars, Washington, D.C., 15–16 June 1987. With the permission of the author, the editor has considerably shortened and edited the original paper for purposes of stylistic clarity.

1. R. Lewin, "Parkinson's Disease: An Environmental Cause?" *Science* 229 (1985): 257–58.
2. Z. Mészáros, "Results of Faunistical Studies in Hungarian Maize Stands," *Acta Phytopath. Hung.* 19 (1984): 65–90.
3. G. Varallyai, L. Rédly, and A. Murányi, "The Influence of Acid Deposition on Soils in Hungary," *Időjárás* (Journal of Hungarian Meteorological Service) 90 (1986): 169–80.
4. Unpublished proposal for the Ministry of Agriculture and Food, Budapest, 1986.
5. Z. Király, "Balancing Chemical and Non-chemical Methods to Manage Plant Diseases, Pests and Weeds," *Proc. Seminar Techn. Sustainable Agric.*, suppl. 34 (Budapest: Agrokémia és Talajtan, 1985): 156–64.
6. A. Gimesi, "Experiments on Weed Control and Defoliation in Sunflower" (in Hungarian), *Növenyvedelem* 19 (1983): 497–501.
7. R. Weidling, "*Trichoderma lignorum* as a Parasite of Other Soil Fungi," *Phytopathology* 22 (1971): 837–45; C. Dennis and J. Webster, "Antagonistic Properties of Species-Groups of Trichoderma: I. Production of Non-Volatile Antibiotics," *Trans. Brit. Mycol. Soc.* 57 (1971): 25–39.
8. L. Vajna, "*Trichoderma* Species and Their Use in the Control of Fungus Diseases of Cultivated Plants," *Növenyvedelem* 20 (1984): 193–201.
9. A. R. T. Sarhan and Z. Király, "Tomatine and Phenol Production Associated with Control of Fusarial Wilt of Tomato by the NO^3-nitrogen, Lime, and Fungicide

Integrated System," *Acta Phytopath. Hung.* 16 (1981): 133–35; A. R. T. Sarhan, B. Barna, and Z. Király, "Effect of Nitrogen Nutrition on *Fusarium* Wilt of Tomato Plants," *Ann. Appl. Biol.* 101 (1982): 242–50.

10. Z. Király, "Plant Disease Resistance as Influenced by Chemical Effects of Nutrients in Fertilizers," in *Fertilizer Use and Plant Health: Proc. 12th Int. Potash Inst. Colloquium in Izmir* (1976): 33–46; B. Barna, A. R. T. Sarhan, and Z. Király, "The Effect of Age of Tomato and Maize Leaves on Resistance to a Non-Specific and a Host Specific Toxin," *Physiol. Plant. Pathol.* 27 (1985): 159–65.

11. B. Barna, A. R. T. Sarhan, and Z. Király, "The Effect of Age."

12. R. D. Schein et al., "Comparison of the Effects of Sublethal Doses of Triadimefon on Those of Rate-Reducing Resistance to *Erysjphe graminis* in Wheat," *Phytopathology* 74 (1984): 452–56.

13. G. Matolcsy et al., "Morphogenetic and Chemosterilant Activity of Asarone Analogues," in *Juvenile Hormone Biochemistry*, G. E. Pratt and G. T. Brooks, eds. (Amsterdam: Elsevier North Holland Biomedical Press, 1981), pp. 393–402; and E. Cohen and J. E. Casida, "Insect Chigin Synthetase as a Biochemical Problem for Insecticidical Compounds," in *Pesticide Chemistry, Human Welfare, and the Environment*, proceedings of the fifth International Congress of Pesticide Chemistry, Kyoto, Japan, 29 August–4 September 1982, J. Miyamoto and P. C. Kearney, eds. (Oxford: Pergamon Press, 1983).

14. C. B. Huffaker, ed., *Biological Control* (New York: Plenum Press, 1971).

15. J. D. Watson, J. Tooze, and D. Kurtz, *Recombinant DNA* (New York: Freeman and Company, 1983); P. Maliga, "Isolation and Characterization of Mutants in Plant Cell Culture," *Ann. Rev. Plant Physiol.* 35 (1985): 519–42.

16. *Mutation Breeding for Disease Resistance Using In-Vitro Culture Techniques*, IAES-TEC-DOC-342 (Vienna: IAEA, 1985).

17. R. D. Schillito et al., "High Efficiency Direct Gene Transfer to Plants," *Biotechnology* 3 (1985): 1099–1103; M. E. Fromm, L. P. Taylor, and V. Walbot, "Stable Transformation of Maize after Gene Transfer by Electroporation," *Nature* 319 (1986): 791–93.

18. A. Meyer, P. Müller, and G. Sembdner, "Air Pollution and Plant Hormones," *Biochem. Physiol. Pflanzen.* 182 (1987): 1–21.

11

SOCIAL SUPPORT FOR ENVIRONMENTAL PROTECTION IN HUNGARY

Miklós Persányi

This chapter addresses the development of an environmental protection policy in Hungary and the social support for this policy. Environmental protection, which initially was regarded as a "passing fancy," today has lasting political significance in Hungarian society. Interest in the environment developed at the beginning of the 1960s and spread quickly, generating intense debate and making many Hungarians recognize that environmental problems would require radical solutions.[1]

ENVIRONMENTAL POLICY: DEVELOPMENT, ACHIEVEMENTS, PROBLEMS

In Hungary as elsewhere, for a long time the state of the natural environment had only a minor effect on the structure and exercise of political power. In recent decades, however, environmental protection has gained enough importance to become a major challenge for the government, and thus an object of political conflict. The belief that it is easy to resolve environmental problems in a centrally planned economy is a misconception.

The state's response to the challenge of environmental deterioration in Hungary has been affected by conflicting interests within society and by contradictions with other priorities. As in other East European countries, the structure and functions of state-directed agencies concerned with the condition of the natural environment have changed and expanded within a relatively short period. In the first stage, the tasks laid out for state agencies were broadened to include sensitivity to the condition of the natural environment. Gradually, environmental protection evolved into an almost separate state function. In the second stage, environmental protection became an autonomous state task with independent national organs responsible for its implementation in most areas. Environmental protection in state policy is constantly evolving, and its goals are not yet fully defined.[2]

Environmental protection is, above all, a responsibility of the state. State organs in Hungary have attempted to develop a body of scientifically based social and economic policies. For example, in 1976 the Hungarian parliament was among the first in Europe to enact a law to protect the environment; more recently, a parliamentary Committee for Habitat Development and Environmental Protection was established. The National Authority for Environmental Protection and Nature Conservation has been set up to manage and coordinate the actions of the various state sectors and to shape the course of environmental protection. Environmental protection also has its own financial chapter in the Seventh Five-Year Plan (1986–90); and the plans of enterprises and local councils include the most important measures to protect the environment. A complex, institutionalized system of control is supported by local committees both in the capital and in the counties.

So far, the state has had some notable achievements in protecting the environment. For example, during the Sixth Five-Year Plan (1981–85), the installation of filters and the closing of some heavily polluting plants led to a decline in the emission of solid industrial air pollutants by 130,000 tons per year, to 30 percent of the 1980 level in 1985. Reliance on less polluting energy resources has also begun to produce a decline in emissions of sulfur dioxide. The long-term decline in cultivated land area has also slowed to two-thirds of its previous rate, and soil conservation has been completed on seven hundred thousand hectares of land. Pollution of groundwater is being brought under control and Hungary's sewage purification capacity has increased by 50 percent, with a current daily capacity of 1.2 million cubic meters. Due to stringent watershed controls, the quality of Lake Balaton's water has also improved.

In the field of nature protection, Hungary's achievements have also received international recognition. About 6 percent of Hungarian territory is protected, with four national parks and 633 protected regions. The populations of several protected animal species have increased. Waste treatment has also improved significantly: one-third of liquid waste from animal breeding plants is being used again. Regulation of hazardous wastes has improved, and communal waste is collected from 60 percent of apartments. Finally, about six hundred plants have been ordered to decrease their noise pollution by a set date.[3] Nationwide research programs have helped to develop technical solutions to environmental deterioration, and nearly 250 new standards have been established to prevent environmental pollution.

Hungary is active in international environmental programs. Especially important are cooperation with other East European countries through the Council for Mutual Economic Assistance and involvement in the United Nations programs. Hungary has signed environmental agreements with Western countries such as Austria, Finland, France, and Sweden, and discussions have taken place with still other countries, including the United States. Hungary is a member of the European air pollution agreement, the Ramsar Convention, and the Bonn and Washington conventions.

The Hungarian authorities have placed much emphasis on environmental education and public awareness as important resources in realizing policy goals. The mass media regularly cover environmental topics.[4] Hundreds of environmental clubs exist in educational institutions, and young people can volunteer or spend time in the three hundred environmental summer camps. Six universities train environmental protection specialists.

The government's achievements in all of these areas demonstrate its commitment to satisfy social needs.[5] The authorities have pointed out, however, that current environmental conditions are not favorable, despite considerable efforts and past successes. Destruction of the environment was not halted during the Sixth Five-Year Plan.

Inadequacies in environmental policy are rooted in several factors: inadequate financial resources; unforeseen, new environmental problems; an imperfect transition from previous policies; and an inconsistent direction. The greatest successes have been in areas where the interests of enterprises correspond to environmental concerns, where wastes and by-products are recycled, and where new technological processes are implemented.

Some pressing environmental problems still confront Hungary. While emissions of solid air pollutants have decreased, acid precipitation resulting from gaseous emissions has decreased the moisture pH. The level of nitrogen oxide has increased in the cities as a result of heavier traffic. Soil fertility is unfavorably influenced by water erosion, and some 2.3 million hectares have been damaged. Pollution of underground water has increased as well. Although traditional pollution of surface waters is declining, microbial pollution is constant or increasing. The supply of running water has increased, but the drainage system has not expanded commensurately. Two-thirds of the volume of sewage water remains biologically unpurified. The nitrate content of water has increased in several settlements. The distribution of green spaces in towns and countryside is uneven. Noise levels are rising in residential areas near roads and railroads. Collection and safe disposal of waste remain a problem, for poorly regulated dumps near larger communities are saturated. Living space for unprotected animal and plant species is decreasing.

During the Seventh Five-Year Plan, however, more resources have been committed to environmental protection. According to the plan, "Environmental protection must be regarded as an important social task, because of its growing importance. Our main concerns are to moderate environmental stress and avoid environmental degradation in order to maintain the potential for improving the quality of life."[6] Managing hazardous wastes, improving the quality of air in heavily polluted areas, protecting water resources both now and in the future, and reducing pollution resulting from insufficient sewage purification are the main priorities. Other unresolved problems also require attention: protecting spas and vacation resorts, developing agricultural technologies, and increasing the effectiveness of nature conservation projects.

Hungary's most important economic task related to environmental protection is to improve economic efficiency and alter the structure of the economy so that the goals of economic development enhance environmental protection. Conversely, the waste of raw materials, energy, and water raises the costs of production, decreases competitiveness, and increases pressure on the environment. The coordination of different state branches and effective involvement of relevant state, political, and social organizations underlie an effective environmental policy. Active participation of both citizens and social organizations is needed to prevent environmental deterioration.

ENVIRONMENTAL AWARENESS AND INTERESTS

Effective environmental protection depends on the active participation of society, for the state cannot make decisions in a vacuum. Even when opinions vary, the voluntary cooperation of individual citizens and organized movements can be constructive in shaping policy and forming public opinion. In Hungary, professionals in the state's environmental organizations believe that spontaneous actions "from below" can be useful if people are familiar with the goals of state policy and can help to protect the environment where they live and work.

Popular input at the local level is also indispensable for aiding official monitoring bodies to identify existing and potential environmental hazards. If people are adequately informed and see that their opinions are valued and their suggestions implemented, they willingly support the state's goals. Citizens' actions have at times played a key role in Hungary, as they have in Western countries.[7]

Environmental pollution does not automatically lead to popular environmental protests in any country. Although excessive pollution may be associated with popular concern over environmental issues, there is no direct link between the two. At least three factors have been found to generate environmental activism among the population: (1) the existence of environmental damage that directly affects the citizens, (2) a relatively high level of satisfaction of material needs in society, and (3) a high level of general social awareness, based on the nature of the political and cultural system.[8] If any one of the three factors is absent, the evolution of an effective environmental protection policy may be hindered. In the long run, however, environmental activism is likely to grow, given the intensification of environmental damage, growing social awareness, and patterns of social development.

Popular activism on environmental issues in socialist countries is both similar to and different from the activism in Western nations. For example, in both types of societies a movement for environmental protection does not start on its own but results from clashes of interest and the ensuing compromises. Likewise, in both types of societies, proponents of environmental protection do not form a homogeneous body but represent different social groups and viewpoints. In both systems there are those who do not approach the problem in its total complexity but make unrealistic demands and resist compromise. Also in both, the environmental movement is, on the whole, critical of existing policy even when the majority of its members are generally well intentioned and constructive.

Without active social involvement, a state's environmental policy cannot be successful, no matter how expertly organized, in either the East or the West. Organized citizen action may go in several different directions, by attempting to influence the population and decision makers either directly or indirectly, or by attempting to alter the state of the environment through volunteer actions.

SOCIAL ORGANIZATIONS AND ENVIRONMENTAL PROTECTION

Views representing various interests in society should reach the political authorities and help shape decisions relating to the environment.[9] From the beginning, popular actions and social organizations have set the tone of environmental protection—a fact recognized at the highest level, the state's environmental documents. According to Hungary's Environmental Act, "The social organizations are to promote the realization of environmental protection goals with their own means."[10] The *Conception and Requirement System for Environmental Protection in Hungary*, ratified by the Council of Ministers in 1980, states, "The active and continuous cooperation of the widest strata of society is an indispensable precondition of successful environmental protection."[11]

Attitudes toward environmental activists in Hungary are as wide-ranging as are attitudes toward environmental protection itself. According to an extreme view that undervalues the importance of environmental protection, the actions of environmentalists have hindered economic development for a long time and present policies are more than enough. The opposite extreme holds that previous and current levels of commitment to environmental protection are negligible. Both extremes, rarely manifested in pure form, are unrealistic and not based on facts. The first view is shortsighted and obsolete; the second, inexperienced. Environmental protection has an important tradition in Hungary. With the worldwide awakening to environmental concerns in the late 1960s and early 1970s, social awareness and activity changed qualitatively in Hungary as well. Hungarian mass organizations and movements played an important role in the early 1970s in formulating the new Environmental Act and later developing a system of state institutions.[12]

Representatives of the scientific community, who were among the first to become involved in international environmental projects, also helped to initiate environmental protection in Hungary. Environmental protection was a focus of attention at the 1971 Annual Assembly of the Hungarian Academy of Sciences, and the following year a special conference was held on the topic. Scientists have also been important catalysts for activity in other organizations. Their participation stimulated the Patriotic People's Front to recognize the importance of environmental problems. In 1973 the front organized its first national environmental protection meeting, the program of which eventually led to the enactment of environmental legislation. The front also undertook the coordination of activities of various environmental protection organizations.[13] It has helped to stimulate public interest in environmental protection, so that now 30 to 40 percent of the comments expressed at meetings of the front are related to the environment. The front has supported the state's environmental measures by influencing public opinion, initiating proposals for changes in decisions detrimental to the environment, and monitoring the quality of the environment.[14]

Young people have been among the most active in promoting environmental protection, and as in many countries, youth organizations in Hungary have been part of the environmental movement from its beginning.[15] The Pioneer Movement has traditionally worked to instill a love of nature through camping, hiking, and tree care. More recently it has launched cleanup actions, revived the "Bird and Tree Days," developed environmental protection guard services, and expanded hobby clubs—all activities with an important educational function.

The Hungarian Young Communist League also has engaged in activities supportive of environmental protection. The forestation movement, in particular, has spread. The league has also participated in planning environmental development, identifying threats to the environment, organizing activities to prevent damage, and providing environmental information.

Young experts joined together to participate in the Youth Environmental Protection Council, the main task of which is to introduce new forms of activity and to support local environmental projects.[16] The council regularly examines development plans that might produce significant environmental damage, in some cases suggesting abandoning potentially harmful projects. For example, in spring 1987 it published its objections to the proposed construction of a pumping reservoir power plant near the planned Gabčikovo-Nagymáros Dam in a nature protection area on the Danube. During a session of the parliament, the deputy from the district where the plant was to be built protested against it.[17]

Labor unions help inform workers and motivate them to protect the environment. Particularly important are their efforts to introduce low-waste technology, recycling, safe disposal of hazardous wastes, and reduced noise and vibration pollution into production processes. Eliminating air and water pollution in the work place is another priority. In the agricultural sector, safe use and storage of pesticides as well as proper disposal of wastes and pollutants are major concerns.[18]

Scientific and technical experts are important participants in the attempts to protect the environment. Specialized associations serve as a meeting ground for engineers responsible for environmental protection, and foresters or biologists. The environmental protection committees of these associations are primarily concerned with improving the production process from an environmental point of view, and they also conduct research and support investments to prevent damage to the environment.

The Society for the Dissemination of Scientific Knowledge also has been active in furthering environmental protection. It disseminates information to help increase awareness about environmental issues and provides a forum for environmentalists to exchange information. Numerous smaller groups are committed to environmental protection. Many of them are made up of "nature lovers": fishermen, hunters, animal protectionists, pet owners, and tourists.

SPECIALIZED CITIZENS' GROUPS

Public opinion is often aroused by specific developments threatening the local environment, such as the reduction of green areas in urban centers and construction in the hilly regions of Budapest. Citizens' groups have often sprung up to protect specific regions, particular green areas, parks, avenues, or trees. Nearly a hundred such groups exist.[19] On a national level, the National League of Town and Village Protection Societies was established in 1986 with the goal of identifying problems and initiating official procedures to prevent harm to nature or to the natural surroundings of settlements. All of these local actions are very important, because even local governments may be insensitive to the environmental impact of their projects. For example, in Keszthely on Lake Balaton, the city council built a road across the park of an eighteenth-century castle in order to divert traffic. Despite a national protest, the council went ahead with the project, insisting on its autonomy.[20] Volunteer social organizations are essential on the local level in residential areas: committees, groups, societies, and clubs deserve strong support.[21]

In Hungary, citizens' groups have also concerned themselves with pollution. In the early 1970s numerous petitions were written concerning air pollution and unbearable noise in Budapest. Citizens have supported the state's efforts to relocate polluting factories. The first such organized citizens' action took place in 1977. As a result of public uproar when children became ill, presumably because of lead pollution, the polluting factory was closed.[22] In 1981, soil and water pollution won nationwide attention when hazardous wastes at the site of a pharmaceutical factory contaminated springs supplying drinking water for the town of Vác; again, the plant was closed.[23] A third serious protest occurred in 1983 when a new runway at the Budapest airport increased noise pollution.[24]

A very heated nationwide debate developed over the importation

of waste from Austria. An enterprise supported by the local council
in Mosonmagyaróvár in western Hungary contracted to dump garbage sent from Graz. The Austrian partner paid with hard currency
and promised to provide high-technology waste management equipment. Chemists from a local environmental group analyzed the waste
and determined that it contained heavy metals and that its pollutants
were reaching groundwater. The group informed the public and the
authorities, leading the National Authority for Environmental Protection and Nature Conservation to stop the importation. Subsequently, a regulation prohibiting the importation of hazardous wastes
was adopted.[25]

Recently, citizens' groups have begun to press for measures to prevent pollution. This kind of protest reflects a higher level of environmental awareness, for it presupposes that the population is informed
about past cases of environmental damage and understands the negative consequences that may result from future actions. For example,
following the incident with hazardous waste at Vác, a strict government decision established waste disposal sites and incineration factories for hazardous wastes. Although these sites were chosen on the
basis of geological research, some of them could not be used because
of opposition from local populations. For example, since 1984 a planned
incineration factory designed to burn dangerous wastes at Dorog has
caused mass consternation. A large part of the town's population
wanted to prevent further air pollution by locating the incinerator
far from residential areas, and two thousand people signed petitions
in protest. The town council backed down after national-level bodies
guaranteed that the plant would operate safely and promised the
necessary financial support to reduce levels of air pollution.[26]

In these cases the population was poorly and belatedly informed
about development plans. When popular protests did draw attention
to serious environmental hazards associated with planned projects,
however, many had a clear impact on policy. Sometimes this impact
involved closing down a factory; other times compromises prevented
damage to the environment. In every case, however, the positive out-
come depended on the active involvement of the population in monitoring environmental conditions and supporting the state's
environmental organizations.

Specialized environmental protection and nature conservation societies are a new phenomenon in Hungary. The Hungarian Orni-
thological Society, founded in 1974 with two thousand members, today
has more than fourteen thousand members. Its work entails protection not only of birds but also of their habitats, and it trains young

people, who make up a majority of its members. Other nature conservation societies have a narrower focus. For example, "Friends of
the Park" groups were formed to support the development and maintenance of national parks.[27] Social organizations that guard the environment have a similar function, but with greater legal authority.
Several county-level nature conservation societies have been formed.

The newest groups, created in the last few years, are more generally
concerned with environmental questions. They are usually formed as
clubs by university students and young graduates to promote stricter
environmental protection policies and change society's values. A good
example of such a group is the efforts of the conservationists' club of
the Budapest Scientific University to protect the Szarsomlyo hill in
southern Hungary. Most of the hill is a protected area of national
importance. On one side of the hill, however, is an old limestone mine
in which mining methods have long endangered the area's wildlife
and geological features and caused much dust and noise pollution in
a nearby village. Many conflicting interests are tied to the mine: those
of the mining industry, environmentalists, and employees of the mine.
The conservationists publicized the conflict and helped to draw up
an environmental impact assessment ordered by the president of the
environmental authority.[28] The compromise that was reached involved a change in the mining technology, thanks to which the fauna
and flora in the area will be protected.

CONCLUSION

Some environmental clubs carry out scientific surveys and help to
expand available information about the environment, relying heavily
on the expertise of their members. But these groups have insufficient
political experience to forge realistic compromises. Nonetheless, this
wide array of environmental protection groups has brought a new
dimension to Hungarian environmental protection. They have helped
along a process by which the environment is becoming an important
question to the whole society. Criticism should not be superficial or
one-sided, but rather sympathetic to the dilemmas facing decision
makers. Environmentalists should be willing to seek realistic solutions
and to propose alternatives. Apart from criticizing existing policy and
warning about potential hazards, social organizations can also play an
important role in increasing the public's awareness of environmental
questions.

Active cooperation between governmental and nongovernmental
organizations is a precondition for effective environmental protection.

Yet environmental activists are only a fraction of the population demanding a sound, harmonious, and healthy environment. The appearance of an active environmental movement in Hungary indicates the rising impatience of the majority of the population both with wasteful and destructive patterns that threaten the environment and with those strata of society that lack appropriate environmental awareness.

NOTES

This chapter is based on a paper presented at the conference "Environmental Problems and Policies in Eastern Europe," sponsored by the East European Program of the Woodrow Wilson International Center for Scholars, Washington, D.C., 15–16 June 1987. With the permission of the author, the editor has considerably shortened and edited the original paper for purposes of stylistic clarity.

1. K. V. Moltke, "Examples of Joint Management of Environmental Education in Europe," *The International Symposium on Long-Term Development of Environmental Policy and Environmental Education in Europe* (Vienna: Austrian Society for the Protection of Nature and the Environment, 1983), pp. 107–14.
2. A. Tamás, "Politika és környezetvédelem," *Állam és Igazgatás*, 1982, no. 12:1074–80.
3. *Az OKTH Elnökének beszámolója az Országgyülés részére* OKTH (Budapest, 1987).
4. I. Kováts and J. Tölgyesi, "Az ökológia kérdései az ujságokban (1982–1983)," *Jel-Kép*, 1986, no. 4:62–70.
5. Gy. Perczel, "Környezetvédelem és gazdaság," *Népszabadság* (April, 1987): 1.
6. "A magyar népgazdaság VII. ötéves terve 1986–1990," *Népszabadság*, 23 December 1985 (supplement).
7. M. Tolba, "Earth Matters," *Environmental Challenges for the 1980s* (Nairobi: UNEP, 1983), p. 21; M. Persányi, "Környezevédelem társadalmi támogatással," *Társadalmi Szemle*, 1986, no. 5:25–38; K. Ábrahám, "A közös gondok közös megoldásáért," *Buvár*, 1985, no. 8:342–43.
8. M. Persányi, "A környezetvédo társadalmi aktivitás feltételei és gyökerei," *Jogtudományi Közlöny* 1987.
9. B. Pokol, "Az érdekképviseleti szervek szerepe a politikai akaratképzésben," *Társadalomtudományi közlemények*, 1985, no. 1:70.
10. "Az 1976. évi II.törvény az emberi környezetvédelemről," *Magyar Közlöny*, 1976, no. 26:318–23.
11. Országos környezetvédelemi koncepció és követelményrenszer, a Minisztertanács 2006/1980./.II.17/Mt.sz.határozata.
12. A. Ádám, "A társadalmi szervezetek szerepe a környezetvédelemben," *Állam és igazgatás*, 1970, no. 7:638–48.
13. Patriotic Popular Front, *Environment and Society: Environment Protection Work at the Patriotic Popular Front* (Budapest, 1979).
14. Patriotic Popular Front, *HNF: III. Országos Környezetvédelmi Konferencia, Gyula* (Budapest: Hazafias Népfront, 1984).
15. A. Peccei, *One Hundred Pages for the Future* (New York: Pergamon Press, 1981).
16. M. Persányi, "Zöldülő generációa," *Ifjusági Szemle*, 1985, no. 4:87–101.
17. T. Korányi, "Tengerszem," *Ötlet*, 12 March 1987, pp. 16–17.
18. "A szakszervezetek továbbra is támogatják a környezetvédelmi programokat," *Népszava*, 17 January 1987, p. 3.
19. A Város-községvédõ és Szépitő Egyesületek Szövetségének Alapszabálya, *Falu*, 1987, no. 1:82–91.

20. A. Mélykuti, "Az elátkozott ut. Az értelem és az érzelem vihara Keszthelyen," *Magyar Nemzet*, 15 November 1986, p. 4.
21. K. Ábrahám, "Környezetvédelem társadalmi összefogással," *Pártélet*, 1986, no. 10: 10–14.
22. B. Fehér, "Az ólomügy története és tanulságai," *Magyar Nemzet*, 2 November 1977, p. 10.
23. István Balogh, *Egy korty halál: Riportok a környezetvédelemről* (Budapest: RTV Minerva, 1982).
24. Gy. Heimer, "Nem tudni, mikor, hogyan sikkadt el a rákoshegyi zajvédelem ügye," *Heti Világgazdaság*, 1984, no. 19:62.
25. V. Nagy I, "Ketyeg az időzitett bomba: Riport a környezetszennyezésről, a felelőtlenségről," *Magyar Nemzet*, 14 February 1987, p. 7; and G. Tarnói and H. Havas, "Szemét ügy," *Mozgó Világ* 1986, no. 4:49–59.
26. László Sólyom, *A társadalom részvétele a környezetvédelemben* (Budapest: OMFB Rendszerelemzési Iroda, 1986); and Pál Támas, *Érdek és kockázatfelismerés* (Budapest: OMFB Rendszerelemzési Iroda, 1986).
27. Z. Rakonczay, ed., *Bükki Nemzeti Park* (Budapest: Mezőgazdasági Kiadó, 1983), pp. 409–10.
28. J. Bunyevácz and J. Tardy, "A Szársomlyó és környékének környezetvédelmi hatásvizsgálata," *Buvár*, 1986, no. 5:8–10.

12

POLISH WATER RESOURCES AND ENVIRONMENTAL PROBLEMS

Jerzy Kurbiel

Almost all of Poland (99.9 percent) lies within the drainage area of the Baltic Sea. The principal Polish river is the Vistula, which is 1,068 kilometers long. Its basin occupies 194,000 square kilometers, 168,000 square kilometers of which lie within the boundaries of Poland and represent 54 percent of the country's area. The second longest river is the Oder, the basin of which covers 118,000 square kilometers, including 106,000 square kilometers within Poland.

The average annual runoff of water amounts to 59 billion cubic meters, equivalent to 31 percent of the country's precipitation. Ninety-three hundred lakes have a surface greater than one hectare, and their total surface area amounts to thirty-two hundred square kilometers, or about 1 percent of the area of Poland. Most of these lakes are in the northern part of the country.

Compared with the rest of Europe, Poland is poor in water resources. Its 1,660 cubic meters of surface water per inhabitant place it twentieth in Europe. The mean unit runoff is 5.5 liters per second per square kilometer of the country's area. Also unfavorable is the irregularity of surface water runoff, which varies from 32 billion cubic meters in a dry year to 90 billion cubic meters in a wet year. There is also significant irregularity between seasons. Only 6 percent of the average annual surface water runoff is stored in impoundment reservoirs drawing on rivers. (By comparison, the degree of impoundment in Czechoslovakia is 12 percent.) Water resources are distributed unevenly throughout the country and frequently do not match the needs in a given area. Extensive economic development, which has resulted in the growth of towns and cities and of new industries, has increased water consumption. In 1961–79, water consumption in the Polish economy grew from 5.6 to 14.2 billion cubic meters per year. Industry uses 71 percent of the water, municipal and rural public supply systems use 18 percent, and agriculture uses the remaining 11 percent. In 1990 the demand for water may reach 32.5 billion cubic meters, or more than half of the average annual stream flow and much more than the available resources of 7 billion cubic meters. Consequently, more and more river water will have to be recycled several times.

POLISH LEGISLATION IN WATER PROTECTION

Principles governing the use and protection of water resources were formulated in 1922, when the Sejm passed Poland's first Water Act. Its provisions remained in force for more than forty years. In response to industrial development and urbanization, the Sejm passed a new Water Law on 30 May 1962, which established certain general principles to protect water resources from pollution:

1. Permission of the water management administration is required to discharge substances that may pollute inland waters.
2. Factories that discharge wastes into waters must construct and operate wastewater treatment plants.
3. Polluters are not only criminally responsible, but they must also compensate for damage.
4. It is forbidden to change industrial processes in ways that may result in increased water pollution.

On the basis of the Water Law, a number of detailed executive regulations have been issued. The most important specifications are as follows:

1. stream standards, establishing acceptable levels of water quality and conditions of wastewater discharge into the water, ground, and municipal sewage systems;
2. principles for setting fines for water polluters;
3. documents required to grant permission for water intake and for wastewater discharge;
4. principles governing the location and technical documentation for investments affecting water quality; and
5. principles for establishing protective zones for water intake and water resources.

The Water Law was amended in 1974. A number of new regulations were put into effect, including fees for using river water and for discharging wastewaters. Income from fees and fines is paid into the Water Management Fund, which finances investments to protect water. The new Water Law also forbids industrial plants to start new production processes before putting appropriate wastewater treatment facilities into operation.

The regulation concerning stream standards specifies a division of all surface inland waters in Poland into three classes of purity according to their planned usage. Class 1 includes drinking water and water for the food-processing and salmonid-hatching industries. Class 2 includes water for animal farming, public baths, recreation sites, and water sports. Class 3 includes water for other industries that do not require water for drinking or for agricultural irrigation. These standards also specify the permissible concentrations of forty-nine physical, chemical, and biological pollutants. The concentration values of typical pollutants are presented in table 12.1. Although legislation has not established effluent standards for wastewater discharges into surface waters, wastewaters are supposed to be treated so as not to exceed the stream standards.

Currently, the control of water pollution is administered by the Ministry of Environmental Protection and Natural Resources, which is responsible for all aspects of the natural environment and management of water resources. Regional administration of water protection is conducted by the forty-nine voivodship-level divisions of environmental control and water management. Each division has its own analytical and environmental monitoring laboratory.

CURRENT STATE OF WATER POLLUTION

Water has become a limiting factor in social and economic development in many regions of Poland. One of the most severe nuisances in everyday life is the shortage of drinking water, which is a result of the gradual departure from established standards for surface water quality. In 1983 the length of rivers carrying water of class 1 purity was less than 10 percent of the total length of Poland's rivers, when it should have been at least 53 percent to satisfy the needs of users. Table 12.2 shows some statistical data on the gradually deteriorating quality of river water, as monitored from 1967 to 1983. In 1985, however, statistical data indicated an improvement in the Vistula and the Oder.

The most polluted river in Poland is the Upper Vistula, including

Table 12.1
PERMISSIBLE CONCENTRATIONS OF POLLUTANTS IN POLAND'S SURFACE WATERS, WITH OTHER PARAMETERS, BY PURITY CLASS

	Purity class		
Parameter	*1*	*2*	*3*
Pollutants (milligrams per cubic decimeter)			
Dissolved oxygen	6	5	4
BOD$_5$	4	8	12
Permanganate COD	10	20	30
COD	40	60	100
Chloride	250	300	400
Sulfate	150	200	250
Dissolved solids	500	1,000	1,200
Total suspended solids	20	30	50
Ammonia nitrogen	1.0	3.0	6.0
Nitrate nitrogen	1.5	7.0	15.0
Organic nitrogen	1.0	2.0	10.0
Total iron	1.0	1.5	2.0
Manganese	0.1	0.3	0.8
Phosphate	0.2	0.5	1.0
Rhodanate	0.02	0.5	1.0
Cyanides (without complex cyanide)	0.01	0.02	0.05
Complex cyanides	1.0	2.0	3.0
Volatile phenols	0.005	0.02	0.05
Detergents	1.0	2.0	3.0
Etheric extract	5.0	15.0	40.0
Lead	0.1	0.1	0.1
Mercury	0.001	0.005	0.01
Copper	0.01	0.01	0.02
Zinc	0.01	0.02	0.02
Cadmium	0.005	0.003	0.1
Chromium +3	0.5	0.5	0.5
Chromium +6	0.05	0.1	0.1
Nickel	1.0	1.0	1.0
Sum of heavy metals	5.0	15.0	40.0
Vanadium	1.0	1.0	1.0
Boron	1.0	1.0	1.0
Arsenic	0.05	0.05	0.2
Fluorine	1.5	1.5	2.0
Acrilonitrile	2.0	2.0	2.0
Caprolactam	1.0	1.0	1.0
Coliform index	1.0	0.1	0.01
Other parameters			
Hardness	7	11	14
Temperature (degrees Celsius)	22	26	26
Color	Natural	Natural	Natural
pH value	6.5–8.0	6.5–9.0	6.0–9.0

its tributaries in Silesia. In the last few years, contamination of these rivers by the discharge of dissolved solids with highly saline coal-mine wastewaters has reached an alarming level. But coal mines are not solely responsible for the Vistula's pollution in Silesia. Municipal, industrial, and agricultural users also contribute greatly to the overall increase of nitrogen, heavy metal, and other pollutants.

Chlorides, total dissolved solids, and ammonia concentration profiles, as observed from 1940 to 1984 at the cross-section upstream from Cracow, provide a good example of the sharp increase in major pollutants in the Vistula. The average annual concentration of chlorides increased more than twenty-five times to an average level of 750 grams per cubic meter in 1983, and 982 grams per cubic meter in 1984. Their maximum annual concentration in these years was 1,400 and 1,380 grams per cubic meter, respectively. But 1983 and 1984 were very dry years, hence the low flow capacity in rivers. The observed minimum annual concentrations nearly approach the concentration limits allowed for class 2 water purity standards that apply to the upstream portion of the Vistula. There are also diurnal variations in chloride concentrations caused by the dewatering of some coal mines.

Table 12.2
WATER QUALITY IN MONITORED POLISH RIVERS,
1967–85

	Percent of monitored river length							Required percent of river length to meet needs
						1985		
Purity class	1967	1973	1976	1979	1983	Vistula basin	Oder basin	
1	31.6	23.4	14.5	9.8	6.8	22.7	11.9	53
2	25.6	32.2	26.7	31.3	28.2	26.7	37.8	40
3	14.0	18.0	24.1	25.9	30.1	16.3	23.0	7
Beyond 3	28.8	26.4	34.7	33.0	34.9	34.3	27.3	0

A significant increase (six times more than the average value) has also been observed in total dissolved solids. Both maximum and average concentrations observed in 1983 exceeded the concentration limits of class 2 and 3 water purity standards. These were 2,655 and 1,667 grams per cubic meter, respectively. Concentrations of ammonia, compared over 1955–84, also indicated an unacceptably high increase (seven times more than the average value). Both maximum (7.8 grams per cubic meter) and average (3.91 grams per cubic meter) annual values exceeded the concentration limits of class 2 standards.

Mercury is a good example of the increasing presence of toxic inorganic micropollutants in the Vistula. Its concentration in all randomly collected water samples exceeds 0.015 grams per cubic meter, when the acceptable limit for class 2 water purity is 0.005 grams per cubic meter. In addition, the concentration of lead (another toxic and hazardous heavy metal), was found to exceed the limit of 0.1 grams per cubic meter, while the maximum lead concentration measured was 0.25 grams per cubic meter.

Based on data for nitrate and phosphate levels from 1977 to 1983, the increase of organic pollutants (albeit not as significant as the increase of inorganic pollutants) is also substantial. Thus the quality of the Vistula's water is being endangered not only by the coal-mining industry but also by other sectors of the economy that discharge toxic, organic, and bioorganic pollutants into the river.

In addition to the extreme case of the Vistula, other rivers supposed to remain within class 1 water quality standards are also polluted. They include the Vistula's tributaries—the Sola, Raba, Dunajec, Wisłoka, San, and Wisłok, all located in the mountainous regions. The Dunajec alone lost 148 kilometers of class 1 water (94 percent of its length) in 1967–77. Its water, though still officially considered of class 1 quality, has had concentrations (in its Tarnów cross-section) of several pollutants much higher than those allowed. Maximum concentrations of these pollutants observed in 1984 were (in grams per cubic meter) ammonia, 4; phenols, 0.04; copper, 0.05; and iron, 1.3. Bacterial contamination with fecal coliform qualified Dunajec water as class 3 quality.

The water of the San, upstream from Dynów and far from industrial sources of pollution, has not met class 3 bacteriological standards for many years. In 1984 the fecal coliform level was 0.001 (instead of 1.0), and the number of bacterial colonies grown on gel at 20 degrees Celsius was 600,000. Other pollutants also occur in concentrations much higher than the class 2 standards. High concentrations of ammonia and detergents (1.2 grams per cubic meter) were detected in 1984; high concentrations of total suspended solids (40 grams per cubic meter at low water levels) and phosphates (0.4 grams per cubic meter) were also found in 1985.

The increased pollution of the Upper Vistula means that farther downstream the quality of the water also deteriorates, threatening municipal and industrial water intakes as well as limiting the river's use for agriculture. The quality of water in Poland's lakes is also unsatisfactory. Of the largest ninety-two lakes with a total volume of 2,726 cubic hectares (or 97 percent of Polish lakes), only 1.8 percent have class 1, 46.6 percent have class 2, and 38.3 percent have class 3 water. The remaining 14 percent have water that does not meet even these standards.

WATER QUALITY PROTECTION POLICY

Protection of water quality must involve both technical measures and improved management, including the following:

1. constructing new, advanced wastewater treatment plants for both municipal and industrial users, with particular attention to eliminating long-overdue wastewater treatment problems in municipal and industrial areas;
2. upgrading existing wastewater treatment plants through more effective operation, modification of conventional treatment processes, and the use of new analytical and monitoring equipment;
3. modifying existing industrial technologies with water-saving production methods, recovering valuable products from wastewaters, recycling water, and building closed-water circuits in industry;
4. recycling water from municipal wastewaters for industrial users in regions with limited water supplies;
5. treating highly brackish wastewater discharges from coal mining using new energy-efficient desalination technologies combined with a controlled discharge of mildly brackish wastewaters to the receiving streams through impoundment reservoirs;
6. controlling runoff from agricultural areas carrying nitrogen, phosphorus, and toxic substances, along with modifications in agricultural and forest management;
7. improving water-quality monitoring systems; and
8. controlling air pollution that may result in soil and water contamination.

SELECTED ISSUES IN WATER QUALITY PROTECTION

A new approach to water management would involve reclaiming water from municipal sewage for industrial and commercial use. This approach would also result in a significant improvement of water quality in the receiving streams. Currently, however, this method is not used in Poland, although the scarcity of usable water in some regions makes it necessary to search for such unconventional sources of water. Water reuse should become an important element of the water management system, especially in southern Poland, Silesia, and the Cracow region.

To resolve the acute water shortage in the Upper Silesian industrial region, water reuse technology has been studied since 1975. The water reuse system that has been developed for this region includes nineteen water reuse plants with a total capacity of nine cubic meters per second. It should be included in future central water planning.

Protection of surface waters from brackish waters has become the most urgent water-related issue in recent years. To protect the Vistula from increased inflow of brackish waters generated by the mining industry, some experts have suggested the following course of action:

1. separating high and low concentrations of brackish wastestreams within the mines;
2. using up highly brackish wastestreams by evaporation (this method has been successful in pilot experiments conducted at the Dębińsko mine treating twenty-eight hundred cubic meters per day of mine water); and
3. transferring mildly brackish wastestreams from Silesia into the Vistula below Cracow by means of a system of transfer lines, pumping stations, and reservoirs with controlled release.

Application of the evaporation technique tested at the Dębińsko mine seems very promising. Both the capital and energy costs of this option are, however, very high. This economic barrier limits the future application of this method even though its expected results are substantial. Evaporation, if implemented, would reduce the total salt-loading discharged into the Vistula by 62 percent and the chloride concentration at the Cracow cross-section from 2,000 grams per cubic meter, the amount anticipated in the year 2000, to 769 grams per cubic meter.

The current state of the Polish economy allows for the construction of only three of twenty planned demineralization plants. For this reason, the projected application of desalination technologies may be feasible only if new methods that use less energy are developed. Furthermore, the transfer of brackish waters into the Vistula, also a very expensive option, raises much controversy and is only a temporary solution. Many countries demineralize water by means of membrane processes, which are very effective but require careful maintenance and close surveillance. In Poland, these processes are still being researched.

In conclusion, the final decision on methods of water reuse and their application in Poland requires further comprehensive planning, research, and technical and economic analysis. The issue of improving the quality of water is closely connected to economic, social, and political considerations.

NOTE

This chapter is based on a paper presented at the conference "Environmental Problems and Policies in Eastern Europe," sponsored by the East European Program of the Woodrow Wilson International Center for Scholars, Washington, D.C., 15–16 June 1987. With the permission of the author, the editor has considerably shortened and edited the original paper for purposes of stylistic clarity.

13

ROMANIA'S ENVIRONMENTAL CRISIS

Edward Mainland

The environmental costs of Romania's forced-draft development strategy since the early 1970s have been severe. Despite the rapidly deteriorating natural environment, the Romanian leadership has not given high priority to public health or environmental protection but has pushed ahead with ambitious chemical, metallurgical, lumbering, and reclamation projects. Pollution has placed increasing stress on public health in virtually all industrial areas of the country. Deforestation of hill regions and sacrifice of wetlands to agriculture have been accompanied by flooding and long-term changes in the water table. According to available statistics, erosion now affects 30 percent of Romania's arable land.[1] The Danube delta, the largest surviving wetland in Europe and the largest expanse of reeds in the world, is threatened by the state's plans to transform most of it into an "agroindustrial complex."

Romania's story casts light on basic questions with which the rest of the planet also struggles. What is "growth"? What is "development"? What is "progress"? Who defines them for society and what are their real costs? What sort of development, if any, is ultimately compatible with a sustainable and healthy society and biosphere? And to what extent may Romania be a special kind of "canary in a coal mine," a harbinger whose form of development, although eccentric, differs only in pace and scale from that of other societies whose elites recoil from Bucharest's methods? Does Romania's environmental record actually point toward a fate that entire industrialized regions of the earth will share, only more slowly? Or are there distinctive features that make Romania special and give its downhill course value as a worst-case scenario?

Romania's postwar economic history is dramatic. Romania is one of Europe's richest natural patrimonies, with abundant arable land, water, timber, and petroleum; it had Europe's fastest rate of economic growth in the 1960s and 1970s; it strove for economic independence from Soviet domination; it reached out for Western capital and technology; and it cherished an ambition to grow still faster. In the early 1970s Romania's leaders debated socialist economic strategy. Moderates opposed accelerationists. Both factions agreed on industrialization, modernization, high growth, and national autonomy in decision making but disagreed on the pace of growth and degree of centralized control. By 1974 the moderates were silenced. Technocratic elements in the Romanian Communist party who argued for science, rationality, openness, and a light hand in cultural policy were pushed aside. President Nicolae Ceauşescu plunged ahead with his enormous gamble on chemicals, metallurgy, and machinery with vast imports of Western equipment, abetted by Western bankers and suppliers. His dream was to make Romania a "developed socialist country" at a level near that of Western Europe in his lifetime.

President Ceauşescu rejected reform of Romania's highly centralized bureaucratized economy and reinforced his own variant of vulgar Marxist ideology. His gamble failed: Romania's uncompetitive products found mostly soft markets abroad. Problems included the economy's inability to adapt commercially to rapid international changes; a work force that had trouble assimilating new technology; professional and managerial cadres plagued by apathy and intimidation; and a president who micromanaged by fiat and whim through family and cronies. By the early 1980s the results were heavy foreign debt, economic retrenchment, internal repression, and austerity. Romania now had Europe's lowest living standard, a condition made even more painful by President Ceauşescu's determination to pay off all foreign debt within a very few years. From 1980 to 1988 Romania's debt fell from $12 billion to $2.5 billion, but at a fearful internal cost. A further cost was Romania's partial return to the USSR's economic fold. In 1985 *The Economist* of London referred to Romania as "Europe's sick man" and "the sick man of Communism."[2]

Romania currently presents an interesting case of reverse development. While growing economically at a statistically impressive pace into the 1980s—both in terms of per capita production of many types of commodities and the introduction of more ambitious forms of technology—Romania was actually regressing, or "undeveloping," along a spectrum of other key indicators: public health, consumer amenities, social cohesion and public morale, individual choice, private housing, heating, lighting, social communication, cultural diversity and vitality, and internal economic and technical innovation. Growth and investment have continued more recently in monumental urban renewal projects and irrigation schemes, but not in sectors with sustainable or renewable export-profit potential. Rather than modernizing, Romania continues to slide backward in many ways as contacts with the outside world become more difficult, sources of foreign information are cut back, and research and development stagnates.

Not unexpectedly, the deterioration of the natural environment, already advanced, has accelerated. A solid, internally written *samizdat* analysis of Romania's backward environmental slide reached the West in 1988, confirming trends noted earlier. Lack of public debate, tight party control over the media, iron commitment to centralized economic decisions, and atomization of society stifled criticism from intellectuals, journalists, and conservationists who might have provided constructive criticism and information about secondary effects. For some time, the regime was generally successful in persuading a large portion of the Romanian public that chemicalized and metallurgical "smokestack development" would bring a better life and that environmental costs were a natural part of modernization everywhere. But "development" merged with the president's personal quest for a place in history and became an increasingly violent and irrational assault on the air, water, and land—justified as "mastering" nature and history through heroic but uneconomical projects such as cutting a canal from the Danube to the Black Sea, sweeping the village peasantry into spartan communal living blocks, and wiping out historic city centers in vast "renewal" projects.

INDICATORS OF ENVIRONMENTAL DAMAGE

By 1983 Romania had risen to fourth place among European countries in per capita emissions of sulfur dioxide. Between 1972 and 1982 Romania recorded Europe's largest increase in these emissions, a remarkable 235 percent.[3] Since then, increases in power plants' use of high-sulfur, low-quality coal have presumably spread still more sulfur dioxide and acid rain. Targeted increases in overall production of brown coal and lignite by 1990 to between 150 and 160 percent of the 1985 level—the highest percentage increase in Eastern Europe— reflect no apparent slowdown or concern about the effects of acid rain.[4] Current forest damage from pollution has presumably gone well beyond the 284,050 acres admitted in 1983—although hardwood forests in Romania may not yet be suffering as conspicuously as coniferous trees farther north, in West Germany.[5]

There have been dangerous health effects from heavy air pollution around major chemical and metallurgical plants, although the impact of pollution on public health has mostly been treated as a state secret and has not been generally publicized.[6] Airborne lead around some nonferrous metallurgical plants, for example, was reported to be fifteen to forty times the acceptable limit.[7] The Romanian press occasionally has noted gardens and forests burnt by sulfur acid deposition and cases of illness—lead, mercury, and cadmium poisoning, as well as pulmonary disorders, rickets, anemia, and tumors in heavily polluted areas.[8] Other effects of pollution, such as the thick, yellowish haze in Hunedoara, acidic enough even a decade ago to blacken and erode the stone walls of that steel town's ancient castle, have been obvious for years. Most industrial units lack air purification equipment entirely or get by with frequently malfunctioning machines of domestic design.[9]

By the early 1980s, 1,364 miles of Romania's rivers were deemed "polluted."[10] Less than 20 percent of the main waterways still provided drinkable water.[11] Local contamination of groundwater—for example, in petroleum-producing areas—had become serious.[12] The Olt, the Danube's largest Romanian tributary and Romania's longest inland river, is described as polluted along virtually its entire 670-kilometer length.[13] Pollution had affected other rivers earlier: in the early 1960s petroleum had so contaminated the lower Siret River that inhabitants accidentally set it on fire, moving an observer to write (with more candor than has been evident in recent years): "Industrialization of the country is moving so quickly that a gap is opening between it and the rhythm of research, design, and execution in the area of water purification. Wildlife losses on the lower Siret stem from this gap."[14] In 1982 more than one-quarter of Romania's 3,801 water purification plants were reportedly still operating below standard, and 68 plants were not operating at all.[15]

Watershed mismanagement has had extremely serious economic consequences. The deforestation of Moldavia, for example, brought floods that impelled authorities to invest in costly dikes, channelization, and impoundments to control the flooding, particularly on the Prut River on Romania's northeastern border with the USSR. Reforestation has been admitted as necessary but usually only as a secondary measure. Romanian experts privately viewed the unprecedentedly intense countrywide floods of 1970 and 1975 as essentially man-made.[16] Water levels changed and the watershed's absorptive capacity was reduced. Modern dams had imperceptibly altered hydrographic configurations, backing up certain interior rivers upstream. Transylvania's pattern of small, centuries-old drainage and catchment systems was shattered by collectivization. Demoralized peasants no longer tended or cleaned these systems and local officials disregarded old ways as traditional communities became fragmented.[17]

The draining of the Danube lakes between Calafat and Brăila had similarly costly consequences. In what once had been one of Europe's richest hunting and fishing areas and a vital stage on the great north-south waterfowl migration route from northern Europe to the Mediterranean and Africa, waterside willow and acacia belts were cut down, watercourses dried up, and farmland of low fertility was created. Such land dries into a kind of dust and henceforth requires large amounts of herbicides, pesticides, and fertilizers to produce crops. The Danube lakes had been natural regulators of climate and water, without which the water table rose up through floodplains and river valleys, causing more frequent floods. Priceless freshwater fisheries meanwhile had vanished.[18]

A similar economic fiasco was the Sadova-Corabia agricultural project in Oltenia (southwestern Romania along the Danube), implemented with considerable help from international lending agencies. Dunes stable for centuries with planted forests and local agriculture were assaulted with intensive irrigation and foreign technological assistance, which turned them into swamps with a clay base that prevented drainage and required much additional investment to recover lost land. Reportedly, misguided irrigation has led to salinization, or the formation of moors, over some two hundred thousand acres of land.[19] Planned "systematization"—first in Ares and Dímboviţa counties (*judete*), then elsewhere—and a future Bucharest-Danube Canal promise more of the same.

THE DANUBE DELTA

Romania's environmental crisis may best be epitomized by the Danube delta, subject of BBC Television's 1984 documentary, "Pelican Delta." Massive development—accelerated reclamation of land for agriculture plus "industrial" fish culture—has now put in jeopardy Europe's greatest wetland, a reserve for rare plants, fish, and birds. As early as 1972, discussions between American scientists and Romanian officials explored the possibility of making the delta a bioreserve for research with foreign funding. At that time, the possibility of hard-currency investments elicited Romania's interest.[20] The Romanian leadership, however, chose the path of massive "capitalization" of the delta's resources. In the 1970s, several failures paved the way. Diked fish ponds produced unmarketable fish and concentrated the Danube's pollutants (notably heavy metals) that had formerly been filtered and dispersed by natural reed beds. The ponds also attracted fish-eating birds: the authorities' violent reaction against them was restrained only by the cost of ammunition.[21] Such alien species as Chinese carp began to supplant and damage native fisheries. Initial cellulose harvesting of reed beds by giant rubber-tired reapers failed economically because overfrequent mechanized cutting killed the reeds (the alternative—infrequent cutting or manual harvesting—did not produce enough cellulose). Plant and equipment losses reportedly reached millions of dollars. Since the mid-1970s, the small institutes and teams of Romanian scientists that study the delta have had trouble obtaining funding because their work generates no immediate value for the economic ministries. These teams have been the one Romanian interest group, struggling however feebly, to lay a basis for saving the delta.

Meanwhile the delta, which fans out eastward from Tulcea into the Black Sea, has experienced "significant pollution from riverside industries dumping copper, mercury, lead, detergents, and oil products." Witnesses report that chemical spills cause "eerie silence" in normally vibrant reed beds.[22] According to one account, "fish species are dying and rare, protected birds who eat the fish are also perishing. The area's mammal population is threatened by development that is destroying its habitat."[23] Once the habitat for 300 species of birds,[24] the delta now harbors only about one-fourth that number.[25] Romania's *Almanacul turistic* (tourist manual) for 1968 listed the delta's area as 500,000 hectares. A 1974 Romanian tourist guide describes "an area of 434,000 hectares."[26] Subsequent sources cite 300,000 hectares or less. Reclamation and development have been nibbling at the edges of the delta for a long time, shrinking natural areas—even though siltation has been extending the delta seaward for millenia at a slow, geological pace, mostly along the Chilia (northern) arm of the Danube, within the Soviet Union.

Current plans for the delta will greatly accelerate diking, damming, filling, and draining. Now 90 percent of the remaining delta appears to be endangered. As much as 144,000 hectares will be "reclaimed for agriculture," old fishing communities will be "urbanized," and 63,000 hectares of industrial fish ponds will be constructed. In contrast, only 1,980 hectares are destined for research, 1,090 hectares for rare species, and 5,930 hectares for bird refuges.[27] Only a maximum of about 45,000 hectares may be "protected" or have their "original character conserved." R. Grimmett, of the International Council for Bird Preservation (ICBP) states that "a much larger proportion of the Delta needs to be protected. The ICBP would support the calls of Romanian ornithologists themselves for a National Park to be created in the Delta."[28] According to Dr. Joost van der Ven of the International Waterfowl Research Bureau (IWRB) in Slimbridge, England, "It is fair to say that some areas in the Delta are now designated as protected areas, but still I am afraid that these little spots will suffer from the hostile surroundings in the future. . . . It is a great pity that we have to watch the destruction of an area that could be developed as Europe's best nature reserve with many different goals and developments."[29] Dr. van der Ven noted that Romanian authorities have been unresponsive to the IWRB's proposals to conduct detailed international waterfowl research in the delta and help Romania avoid extirpation of the small number of remaining endangered pelicans and the dwindling number of swans, ibises, herons, egrets, birds of prey, and waterfowl inhabiting reed beds and lakes. Hundreds of thousands of geese, storks, ducks, and wading birds migrate through the delta in the spring and fall. The ICBP has called attention to the delta's large size and crucial location on the Black Sea coast, an irreplaceable wetland like no other for these migrant species. Romania is an official member of the IWRB but so far has refrained from signing the 1971 Convention on Wetlands of International Importance (the Ramsar Convention) by which some forty signatories, including most of Romania's neighbors, seek to protect exactly what the delta and other similar sites contain. Lawrence Mason of the U.S. Fish and Wildlife Service notes the "extraordinary productivity" of estuarine areas such as the delta, which produce more than twice the organic material of adjacent croplands and provide vital spawning, nursing, and feeding grounds for fish. "An island here and a corner there will not maintain this biological productivity," according to Mason.[30]

Major new canals are being sliced through the delta to straighten and channelize the Sfîntu-Gheorghe arm of the Danube into the Black Sea. Belatedly, some Romanian experts recognize that accelerated water flow will radically alter the pattern of alluvial deposits and recreate problems of siltation, change in water temperature, salinity, fog, and ecological imbalance that were evident after the construction of Romania's Danube–Black Sea Canal.[31] Another potential threat to the delta, particularly in the wake of the Chernobyl accident, is the current construction of Romania's first nuclear power plant, upstream along the Lower Danube at Cernavoda. The plant has been plagued by delays, incompetence, political interference, poor supervision, and unrealistic targets. Just south of the delta and north of Constanța, at

Navodari, a large petrochemical complex is slated for expansion—
foreign capital willing—that may well blow heavy pollution either
north into the delta or south over Romania's Black Sea beach resorts,
a profitable source of foreign exchange in recent decades.

· It remains to be seen whether Romania will be receptive to any new
foreign initiative to protect remaining areas of the delta as, for example, in the "debt for nature" transaction that Conservation International concluded with Bolivia in 1987. This nonprofit U.S. group
retired $650,000 of Bolivia's foreign debt in return for a 3.7-million-
acre expansion of conservation areas in Bolivia.[32] A comparable action
might, under certain circumstances, look advantageous to Romanian
authorities.

LESSONS OF ROMANIAN DEVELOPMENT

Can object lessons be drawn from Romania's environmental crisis?
The following are some conclusions that might be drawn from the
Romanian experience.

1. *Unimplemented, unenforced national laws may be as bad as no laws at
all.* The Romanian government enacted comprehensive legislation in
June 1973—the Environmental Protection Law, which has had little
discernible effect and has not been enforced in subsequent years. As
in many countries, Romania's state agencies themselves widely ignore
or circumvent environmental laws, which provide a cosmetic screen
for deliberate inaction.

Stifled media and intimidated public opinion do not provide the
government with the feedback and professional advice necessary to
consider all aspects of these problems and to arrive at sound decisions
to avert environmental catastrophes and economic waste. The suppressing of divergent opinions, especially at the highest level, explains
in part the ongoing ruination of Romania's environmental, social, and
economic fabric. Failures should help to reveal future paths to success,
but in Romania failures have been repeated without lessons being
learned. For example, the draining of the Danube lakes is being followed by the draining of the Danube delta.

Control of public discussion has also engendered public apathy
toward severe environmental damage, ironic in a country whose popular myths are rooted in nature. A Romanian proverb synonymous
with national identity, *"Codrul frate cu românul,"* says, "The forest is
the Romanian's brother." Forests once were home to Latin-speaking
proto-Romanians hiding to survive the wave of barbarians who swept

across the Balkan territories of the late Roman Empire. The national folk ballad "Miorița" also is colored by nature mysticism. Such folk roots were formerly a deterrent against environmental mischief.

2. *Growth does not necessarily mean development.* By conventional economic indexing, Romania has unquestionably "grown" in some sense. A large number of "white elephant" industrial plants throughout the country have the capacity to produce a greater volume and diversity of products than before. Monumental urban construction and unrealistic irrigation schemes have caused visible change. Yet the society's capacity for sustained indigenous economic or technical innovation is failing, and the quality of life in Romania has deteriorated seriously according to most indexes used in the West. Not only have ordinary amenities—food staples, heat, light, appliances, medicines, entertainment—become luxuries or unknown items, the texture of social life has become harsher, gloomier, and more simplistic.

Conventional development based on heavy industry may not be the only path to general economic progress. In fact, on the threshold of the twenty-first century, building more industry that pollutes and poisons may prove to be the precise reverse of development. Were it not for the incubus of Stalinist economics, there would seem to be no theoretical reason why postwar Romania could not have achieved substantial prosperity by unleashing its great potential and existing comparative advantage in natural resources and agriculture along noncollectivized lines, allowing this prosperity to generate a balanced, selective, specialized industrialization. After all, no one denies that Denmark and Finland are "developed." Countries of the Pacific Rim have marked out yet another special path to progress.[33]

Furthermore, conventional economic reckoning fails to embrace the quality of life or environmental costs. Conventional economic accounting has conveyed a misleading statistical picture of what really has been happening in Romania during the past several decades. From the viewpoint of bioeconomics, Romania has been quite rapidly moving from a low entropic condition to a high entropic condition.[34]

3. *International methods of environmental protection break down when the will of national authorities is stronger than that of international guardians and covenant keepers.* There has been a dearth of international concern about the Danube delta and other environmental problems in Romania, even though they have reached crisis proportions at a time when foreign critics of Romania's human rights record have been vocal and legion. Aside from the BBC documentary "Pelican Delta," few comments have come from private environmental organizations, and little testimony from distinguished Western ornithologists who

worked in the delta in happier times. Nor was there any explicit directive about the delta in the debates at the Wetland Convention at Regina, Saskatchewan, from 27 May to 5 June 1987, although the delta is a valuable international resource.

International lenders need watchers and a "sunshine rule" (openness and full disclosure) to keep them from funding environmentally harmful projects that make no economic sense. Harm has been done in Romania through the financing of projects by international agencies and multinational banks. Environmental impact has been considered only formalistically, if at all, in many instances. For example, the World Bank's recommendation to fund the diversion of water from the last free-flowing river in Romania's Southern Carpathians (Riul Mare, south of Hunedoara) in the mid-1970s was accompanied by an environmental statement that in effect contended that there would be no environmental impact, even though virtually the entire stream flow—along the border of what was formerly Romania's most sacrosanct mountain scientific reserve, Retezat—was to be drained out of its bed toward new chemical plants.[35]

Romania's government imposed a quasi—"hermit kingdom" status on its people during the 1980s to achieve discipline and order under conditions of austerity. Yet mounting pollution has ignored state boundaries and is being felt in neighboring countries. Environmental solutions must transcend political frontiers. The flow of ideas and information across borders may help people to find ways to survive and renew national resilience in a manner analogous to biologically regenerative coevolution and ecological interdependence among organisms.

When speaking of a "crisis," one usually has in mind either a turning point or a special time of danger for a definite domain of human, social, or national interest. According to all evidence, such is the situation in Romania at this time and for the immediate future. The downhill slide of environmental deterioration has reached the point where many Romanians think that nature may never recover. Yet is Romania's condition so different from the world outside its borders? There, despite several decades of environmental awareness, legislation, and eloquence, the great poisoning, crowding, bulldozing, and impoverishment of the biosphere continue. The liberal, affluent, modernized, industrial democracies have failed, despite intermittent success, to stem a whole range of seemingly implacable trends: dying estuaries, wilting or mutilated forests, disappearing species, shattered habitats, poisoned rivers, ebbing aquifers, and overflowing wastes. Doubt therefore remains, notwithstanding a high-level United Nations-sponsored study to the contrary, about whether a healthy biosphere and conventional industrial development can coexist.[36]

Recovery from ecological blows requires tolerance of contending viewpoints, a general civility between governments and governed, a capacity for creative initiative, and a broad redefining of modernization, growth, and progress. Romania, like the rest of us, first needs to remember what it has forgotten and then radically to refocus its vision of what is necessary to survive.

NOTES

1. *Radio Free Europe Background Report*, no. 42, 20 March 1987, p. 21.
2. *The Economist*, 26 October 1985, p. 15.
3. *Radio Free Europe Situation Report*, no. 19, 29 January 1983, p. 18.
4. See the chapter by John Kramer in this volume.
5. Christine L. Zvosec, "Environmental Deterioration in Eastern Europe," *World Affairs* 147, no. 2 (fall 1984): 101–2.
6. Ibid., p. 102.
7. Ibid., p. 101.
8. Ibid., p. 102.
9. Ibid.
10. Ibid.
11. John M. Kramer, "The Environmental Crisis in Eastern Europe: The Price for Progress," *Slavic Review* (summer 1983): 204.
12. Zvosec, *World Affairs*, p. 102.
13. *Radio Free Europe Background Report*, no. 42, 20 March 1987, p. 21.
14. N. Bacalbasa, "Pluarea zonei inferioare a riului Siret si urmarile primejdioase pentru fauna acvatica" (Pollution of the Lower Siret River and Dangerous Consequences for Aquatic Fauna), *Ocrotirea naturii* (*Nature Protection*), no. 2 (1966). Many other informative articles on environmental questions could be found in this publication before the authorities joined it with two other journals to form a new journal commonly called *Terra* (the full title is *Ocrotirea, mediiului, inconjurator, natura, terra* or *Protection of the environment, nature, terra*). According to a Romanian editor the change was made lest people get ideas about taking environmental protection too seriously.
15. *Radio Free Europe Situation Report*, no. 2, 21 November 1983, p. 18.
16. Personal conversations with Romanian conservationists.
17. Ibid.
18. According to private Romanian Communist party sources, the draining of the Danube lakes appeared to have been originally inspired by the example of the Virgin Lands campaign of Nikita Khrushchev, the deposed Soviet party leader whose successors criticized him for "hare-brained scheming" and "voluntarism."
19. *Radio Free Europe Background Report*, no. 42, 20 March 1987, p. 21.
20. Conversation with a Romanian diplomat present at the 1972 talks.
21. According to witnesses, the authorities hired cadres of armed guards to shoot on sight such fish-eating birds as ibises and herons. The campaigners described such birds as "enemies" of the state's fishing industry (the ibis eats mainly frogs and the pelican less than one kilo of fish per day).
22. Personal conversation with Romanian conservationists.
23. Zvosec, *World Affairs*, p. 102.
24. Sebastian Bonifaciu, Nicolai Docsanescu, and Ioana Vasiliu-Ciotoiu, *Romania, Tourist Guide* (Bucharest: The Publishing House of Tourism, 1974) (also Freiburg, West Germany: Rombach Publishers, 1974), p. 332.

25. U.S. Fish and Wildlife Service, private communication (August 1987).
26. Bonifaciu et al., *Romania*, p. 328.
27. Constantin Acmoja, "Largi perspective de desvoltare turismului in Delta Dunarii" (Broad Prospects for Developing Tourism in the Danube Delta), *Informatia Bucurestiului*, 23 June 1987, p. 1.
28. R. Grimmett, ICBP, in radio interview by Dorin Tudoran, Romanian Service, Voice of America, 10 August 1987.
29. Letter to the author, 28 July 1987.
30. Lawrence Mason, U.S. Fish and Wildlife Service, Washington, D.C., in radio interview by Dorin Tudoran, Romanian Service, Voice of America, 10 August 1987.
31. *Radio Free Europe Situation Report*, no. 13, 26 November 1986, p. 37.
32. *New York Times*, 14 July 1987.
33. Nicholas Georgescu-Roegen, *The Entropy Law and the Economic Process* (Cambridge: Harvard University Press, 1971). See especially the discussion of the "fallacy of the industrial axiom," p. 329, within a general critique of the inadequacies of standard economics, and of the failure of Marxist and non-Marxist schools to inquire into the "process by which new *economic* means, new *economic* relations are created," pp. 316–30.
34. Nicholas Georgescu-Roegen, "Energy and Economic Myths," *Southern Economic Journal* 41 (January 1975): 347–81. See especially the introduction, where Georgescu-Roegen discusses conceptual, mechanistic, tautological, and epistemological problems that weaken the analytic usefulness and ecological relevance of economics.
35. The author personally visited the Riul Mare basin in 1975 and read the documents cited.
36. See the World Commission on Environment and Development, *Our Common Future* (Oxford and New York: Oxford University Press, 1987).

14

YUGOSLAVIA'S ENERGY CHOICES AND THE ECONOMIC DIMENSION

Mihailo Crnobrnja

It is not an easy task to survey Yugoslavia's numerous and interrelated current problems related to energy, the economy, and the environment. In this chapter, these problems are put in a broader context. Rather than review the whole range of issues, I focus on what I see as the most important choices lying before Yugoslavia. What follows should be viewed as a rough sketch rather than a detailed account.

THE BROADER ECONOMIC AND POLITICAL SETTING

For some time now Yugoslavia has suffered from serious economic problems and a high degree of economic instability. On the surface, the problems take the form of rampant inflation, severe unemployment, profound regional disparities in the level and dynamics of economic activity, and a relatively large external debt. The rate of growth of the gross national product (GNP), which for several decades was among the highest in the world, fell to a bare 1 percent in 1980–85. Underlying these problems are deeper inefficiencies in the economic system.

In the mid-1970s an attempt was made to further the expansion of self-management, the hallmark of the Yugoslav economic and political system. The importance of markets, laws of supply and demand, cost and benefit structures, and competition was reduced, while the system of social contracts (or compacts) and self-management agreements was expanded. These institutional changes were later dubbed "contractual economics." The whole idea was to substitute contract and negotiation for competition and coercion among economic agents.

The new methods did not work as expected. Objective market criteria were undermined. As the political authorities sought to demonstrate the feasibility of contractual economics, they increased government involvement in economic matters, presumably on a temporary basis. But government intervention turned into almost outright regulation of most economic decisions, which were originally supposed to be made by decentralized economic units (primarily self-managed enterprises).

These centrifugal forces within the economy, coupled with political decentralization of the Yugoslav federation, produced a tendency in all levels of government to hide the material imbalances that arose from unsound economic behavior. The result was a debt to foreign creditors of almost $20 billion and an accumulation of assorted internal debts.

A secondary result, more relevant to our concerns here, was a very lenient energy policy involving relatively high levels of energy consumption per unit of produced output. The environment fared even worse. The factors of production had no market prices or only seriously distorted ones, and accurate cost-benefit analyses could not be made. Environmental costs, which are very difficult to internalize, were therefore not given appropriate weight. Environmental protection was not institutionalized during this period, despite the passage of numerous laws, bylaws, and regulations; when institutionalization did happen, it was incidental and not systematic.

In 1983 a turnabout was attempted. The country adopted the Long-Term Program of Economic Stabilization (LTPES), which was to be a renewed strategy of greater reliance on market guidance of economic decision making. Setting prices at the world market level was considered important, for it introduced an external, and therefore objective, basis for decisions in Yugoslavia. In the four years since the inauguration of LTPES, very little has been accomplished along the strategic path established by the program. The momentum of LTPES slowed considerably in 1984 and 1985 and was somewhat revived in 1986. In setting the broader background for the discussion of energy policy in Yugoslavia, one must consider the vast differences that exist among regions in terms of levels of economic development on the one hand, and energy endowment and production on the other. The more developed republics of Slovenia and Croatia are short of energy supplies, while the less developed republics and provinces of Bosnia, Herzegovina, Serbia, and Kosovo have much greater energy resources. Because the exploitation of energy is very expensive and capital intensive, the less developed regions have little incentive to develop their resources, whereas the more developed republics and the province of Vojvodina show some reluctance to invest in energy production outside their own territories.

Coordination of investments in the energy sector could potentially bring savings to all concerned, particularly in the power and coal subsectors. In principle, closer coordination would benefit both the "importing" republics and the autonomous province of Vojvodina, which would obtain energy at a lower cost, while also helping the "exporting" republics and the other autonomous province of Kosovo, which would gain revenues from the sales of energy.

ENERGY AND ECONOMICS IN THE PAST

Energy policy in Yugoslavia is at a crossroads. The principal task is to adopt policies to reduce the waste of energy in the production process and to select a more rational energy mix. Yugoslavia is not rich in resources; it is below the world average in energy endowment and just slightly above the European average.

The development of energy has been a priority in the past four decades. Investments in the energy sector represented between 8 and 15 percent of all investments during this period and up to one-third of industrial investments. For example, in 1983, total energy-sector investment amounted to $1.15 million. One might well conclude that these investments were too high, unnecessarily raising the real economic cost of energy. At the same time, the consumption of energy was expanded by setting the price of energy unrealistically low.

Investments in the energy sector were made primarily on the premise that energy production is the locomotive of growth. This assumption was so powerful that the main expansion of liquid-fuel power plants occurred *after* the first oil shock of 1973. It was generally believed that even the high oil prices could be efficiently integrated into production costs. This approach turned into one of the sources of cost-push inflation that still prevails in Yugoslavia.

The level of energy consumption in Yugoslavia is very low by European standards—only Greece and Portugal are lower. But in terms of total energy use and electricity consumption, Yugoslavia is about half as efficient as the European countries of the Organization for Economic Cooperation and Development (OECD) are on average.

Yugoslavia is by far the poorest country in this group in terms of energy provision for its inhabitants, but also by far the least efficient in the conversion of energy into gross domestic product. Therefore there is considerable scope for improving efficiency; at the same time, Yugoslavia must expand its production and consumption of energy.

Currently Yugoslavia is undergoing a reexamination of priorities in the energy sector, based on a critical review of past successes and failures and an identification of relevant constraints, costs, trade-offs, and policy options.

ENERGY AND ECONOMICS: CURRENT PROBLEMS AND ISSUES

Pricing is an important starting point and the first major issue for a more adequate and rational energy policy particularly since Yugoslavia's market-oriented reforms require economic units to make their decisions on the basis of market information.

The declared government policy is to remove disparities between energy prices and other prices, that is, to raise the relative price of energy and increase pressure for its rational use. At the same time, current policy calls for removal of disparities between prices of energy from different sources such as imported oil and domestically produced electricity from hydropower and thermal plants.

By 1988 electricity prices are to be adjusted to target levels based on so-called common elements of pricing. These common elements consist of (1) average financial cost of electricity production in Yugoslavia, (2) world level-relative prices, and (3) relative prices of energy products. According to this scheme, the federal government would maintain responsibility for setting prices of crude oil and petroleum products, while prices of other energy products would be set by the enterprises in the various energy subsectors.

There is still controversy in Yugoslavia over whether this is the proper way to determine realistic energy prices. Although implementation of the policy would improve the financial position of energy producers (the producing federal units), it remains disputed because the common elements of pricing are based mainly on financial, not economic, criteria and may require serious adjustment when prices begin to reflect relative costs.

The common elements for coal pricing include components (about 40 percent) that reflect prices of internationally traded coal, natural gas, and heating oils, adjusted for calorific values of specific types of domestic nontradable coal. These components have little relevance to the resource costs of domestic lignite and brown-coal production, which makes up more than 90 percent of total coal production in Yugoslavia. Allowing international pricing of energy resources to affect domestic coal prices risks making coal prices too high, which would (1) affect the competitiveness of the Yugoslav industry and (2) in the light of persistent inflation, lead to higher imports of energy. Furthermore, although the common elements are useful in the short term for adjusting prices to meet the financial requirements of producing enterprises, they provide little information on the real economic costs of supply. This information in turn is important for achieving efficiency in both operations and investment selection. In short, it is critical for the strategic goal of the policy.

Investment policy in the energy sector is the second major issue. The situation is complicated by the fact that investment policy includes decisions not only about how to produce energy but also about how to produce the equipment needed to produce the desired energy mix. For example, Yugoslavia today has the technical capability and know-how to produce an entire hydropower plant, 50 to 60 percent of a coal-powered thermal plant, and about 25 percent of a nuclear power plant. Consequently, decisions favoring one type of power over another automatically have implications for the use of domestic or imported equipment.

Currently there is considerable excess capacity (about 10 percent) in liquid-powered thermal plants, which is grossly underused because of shortage of hard currency to purchase oil. The cost and availability of foreign exchange is one of the biggest problems currently confronting Yugoslavia. Often, however, the choice is between using foreign exchange to purchase oil for existing power plants and using it to purchase machinery and equipment for the construction of new power plants. Yugoslavia's unstable macroeconomic environment and foreign debt problems have rendered the energy sector unable to attract foreign commercial credits, which previously constituted a significant portion of foreign currency. Only if most of Yugoslavia's additional power in the next decade or so comes from hydroplants will this dilemma be avoided.

During the peak of political decentralization in the 1970s, the republics and provinces of Yugoslavia implemented energy investment programs designed basically to meet their internal objectives of self-sufficiency. These programs were only loosely coordinated and from a national perspective often did not represent the most cost-effective approach. In addition, they were severely delayed, mainly as a result of financial shortages. As a result, Yugoslavia imported the same share of its total energy supply in 1985 as it had in 1973—about 35 percent—despite the policy goal of substituting domestically produced energy for imports.

The current Five-Year Plan (1986–90) calls for a nationwide development policy for the energy sector, including common planning of energy production, consumption, and construction. The goal is a single energy system and a single system of pricing for the whole country. Standing in the way of this attempted rationalization of energy investment are both technical problems and powerful vested interests in the six republics and two autonomous provinces. If closer coordination and planning of investment outlays in this capital-intensive sector are to succeed, enterprises must have reference to a meaningful interest rate, be able to obtain foreign exchange without resorting to uneconomic (and often illegal) means, and be held accountable for financial losses. Even if these problems are solved, investment in energy production may still be economically and politically difficult, because (1) energy investment creates few new jobs directly and (2) most of the one million unemployed are in the less developed republics and provinces—which are also more richly endowed with energy resources.

There is considerable scope for energy conservation in Yugoslavia. It is especially important in the industrial sector, which accounts for the greatest share of energy consumption. Energy conservation is one component of the effort to promote efficiency in the economy. For large industries that have unused capacity, are uncompetitive, or have inappropriate technology, energy conservation must become part of a program to restructure the industrial subsector. In the past few years a number of industrial subsector studies have been proposed or initiated. The cement, chemical, and steel subsectors are examining the interrelated issues of restructuring and energy conservation. It would, of course, be desirable to carry out such studies in other capital- and/or energy-intensive industries, for example nonferrous metallurgy and petroleum refining.

The fall in international oil prices in 1986 called for a reevaluation of priorities in energy conservation, or at least for setting up a different time frame for their execution. It is estimated that the Yugoslav energy import bill declined by $800 million to $1 billion in 1986. This windfall, of course, delays the pressure for conservation and increased efficiency but does not permanently abolish the need for it.

ECONOMIC FACTORS AFFECTING THE NUCLEAR OPTION

Currently, a heated debate is going on in Yugoslavia regarding the projected expansion of nuclear energy. At present Yugoslavia has

Table 14.1
PROJECTED NUCLEAR ENERGY DEVELOPMENT IN
YUGOSLAVIA
(PERCENT OF TOTAL ELECTRICAL PRODUCTION),
1985–2020

Year	*Percent*			
	Coal	*Hydro*	*Nuclear*	*Other*
1985	44.2–45.5	42.7–41.4	5.2–5.1	7.9–8.0
1995	47.1–53.5	40.2–35.0	8.0–7.3	4.7–4.2
2010	57.0–46.2	25.1–19.8	14.0–30.2	3.9–3.8
2020	40.9–29.9	17.3–13.3	38.0–53.1	3.8–3.7

SOURCE: LTPES, 1983.

only one nuclear power plant, which in 1985 supplied 5.2 percent of total electrical production. Environmental and safety issues are also of primary concern in the public debate. Here, however, we will focus on economic considerations.

The LTPES included a section titled "The Long-Run Strategy of Energy Development in Yugoslavia," calling for an additional three to four nuclear power plants by the end of this century, and between six and sixteen more by the year 2010. By the year 2020, nuclear power was supposed to constitute about one-fifth of primary energy sources. (In terms of electricity alone, the projections were those shown in table 14.1.)

One objection to these projections centers on disputable estimates of future energy needs. Many knowledgeable economists maintain that future energy requirements have been "inflated." The projections underlying the long-run energy strategy represent a simple correlation coefficient of GNP growth and energy requirements. According to these projections, electrical consumption would increase almost sixfold in thirty-five years (table 14.2).

An alternative approach is to view energy requirements as a decreasing function of the growth of the GNP. According to this hypothesis, the relevant energy requirements would be considerably lower (table 14.3). If this second set of projections proves feasible, nuclear power generation would become unnecessary until well into the twenty-first century. With proper economic policies and pricing and investment decisions, increases in energy consumption would be slowed, thus postponing the inevitable large-scale introduction of nuclear power.

Table 14.2
PROJECTED GROWTH IN ENERGY CONSUMPTION IN YUGOSLAVIA (LINEAR CORRELATION METHOD), 1985–2020

Year	Energy requirement (billion kilowatt hours)
1985	75
1990	98
1995	134
2000	176
2010	282
2020	425

SOURCE: LTPES, 1983.

A second objection to rapid increases in nuclear power production stems from the present severe shortage in investible funds. A nuclear power plant is, on average, 20 to 30 percent more expensive than a coal-powered plant. The trade-off between initial capital outlays (which are higher for nuclear power) and operating costs (which are lower for nuclear power) is particularly significant in this context. Today, the capital outlays for nuclear plants loom prohibitively large. Each nuclear power plant (nine hundred megawatts) costs about $2.5 bil-

Table 14.3
PROJECTED GROWTH IN ENERGY CONSUMPTION IN YUGOSLAVIA (NONLINEAR CORRELATION METHOD), 1985–2020

Year	Energy requirement (billion kWh)[a]	Difference from linear correlation method (billion kWh)
1985	75	—
1990	90	8
1995	110	24
2000	131	45
2010	152	130
2020	176	249

SOURCE: LTPES.
[a]kWh = kilowatt hours.

lion; thus the outlay would be between $15 billion and $40 billion for nuclear power plants in the next thirty-five years.

Foreign exchange and foreign debt problems further complicate the situation. Construction of nuclear power plants would require much foreign equipment and thus demand outlays in foreign currency. On the other hand, thermal coal and hydrogeneration could be predominantly equipped and financed by domestic resources. The nuclear option would inevitably lead to a doubling of foreign debt.

ENVIRONMENTAL IMPACT OF VARIOUS ENERGY OPTIONS

Given the bad state of the Yugoslav economy, investments in environmental protection have been under great pressure. Enterprises are expected to modernize in order to improve their competitiveness, and at the same time to invest their own resources in nonpolluting processes or waste purification. This expectation may sound like an ordinary application of the "polluter pays" principle, because there are no public subsidies for environmental protection. But imposed and (in the case of energy) below-cost prices work against the environment, adding to the depletion of natural resources. The pricing system impairs the polluter-pays principle and makes it acutely difficult for enterprises to invest in environmental protection.

Given these circumstances, the courts are generally lenient toward enterprises that violate environmental regulations. Even financially sound enterprises often face conflicts regarding environmental expenditures. Confronted by a declining standard of living, workers in the system of self-management may find themselves having to choose between investing resources in environmental improvements or allocating funds to the payroll.

The currently weak economy produces one set of obstacles to environmental protection. But the polluter-pays principle introduces another problem. If a calamity occurs and the enterprise responsible for it cannot afford to clean up the pollution or goes into liquidation, no alternative source of funds is available. There is no insurance system, and the allocation of large sums of government money for these purposes seems highly unlikely. At the same time, there is no built-in incentive for firms to anticipate the possibility and prevent environmental damage. Given these problems, the most effective environmental protection in Yugoslavia is reliance on sources of energy that minimize environmental hazards.

Hydropower is a major domestic source of electricity. Despite shortages resulting from limited rainfall in recent years, further expansion (possibly even a doubling) of hydroelectric power generation is a high priority in Yugoslavia's energy policy. Currently only about 40 percent of the hydropotential is being exploited, while the comparable figure in most countries of the Organization for Economic Cooperation and Development that have high hydropotential reaches 90 percent. Expanding hydroelectric power generation introduces the fewest environmental costs and hazards. Therefore, the immediate priorities in energy expansion at the same time benefit the environment.

The greatest potential environmental threats confronting Yugoslavia in the next few decades arise from the planned increase in the exploitation and use of domestic lignite for thermal power generation. Open-pit mining damages the landscape and creates problems for mine overburden disposal and reclamation of derelict land resulting from past mining practices; furthermore, burning low-calorie lignite with high ash and sulfur contents produces major air and water pollution.

Organizational and political decision-making structures in Yugoslavia may hinder environmentally sound policy making in the energy sector. Most decisions concerning large-scale energy projects are made at the republic and province levels and involve little, if any, interaction with the local authorities. On the other hand, environmental planning (a component of physical planning) is almost entirely the responsibility of individual republics and is often delegated to the local communes. Yugoslavia therefore faces an organizational problem necessitating careful allocation of responsibilities and cooperation among all responsible parties involved in economic, energy, and environmental planning at the federal, republic, and commune levels.

NOTE

This chapter is based on a paper presented at the conference "Environmental Problems and Policies in Eastern Europe," sponsored by the East European Program of the Woodrow Wilson International Center for Scholars, Washington, D.C., 15–16 June 1987. With the permission of the author, the editor has considerably shortened and edited the original paper for purposes of stylistic clarity.

ABOUT THE AUTHORS

Mihailo Crnobrnja, who received his M.A. from the University of Maryland and his Ph.D. from the University of Belgrade, is currently minister of economic planning of the Socialist Republic of Serbia in Yugoslavia. He is also an associate professor at the University of Belgrade, teaching courses on economic policy and political economy. He is author of two books and numerous articles and coauthor of two other books.

Imrich Daubner, who received his Ph.D. in science at the Slovak Academy of Sciences in Bratislava, is director of the Institute for Experimental Biology and Ecology at the academy, specializing in water pollution. He is a member of the Czechoslovak Academy of Sciences and serves as president of the International Association for Danube Research in Vienna. He has written *Microbiology of Water* and *Membrane Filters in the Microbiology of Water* and some 150 articles and scientific works, most recently "On the Question of Microbial Indications of Water Quality."

Joan DeBardeleben is associate professor of political science at McGill University in Montreal. She received her Ph.D. from the University of Wisconsin in 1979. She is author of *The Environment and Marxism-Leninism: Soviet and East German Perspectives* and "Political Legitimation and Economic Policy: Natural Resource Pricing in the USSR and GDR" and coauthor of *European Politics in Transition*.

Georgi Gergov is a senior researcher at the Institute of Meteorology and Hydrology in Sofia.

Barbara Jancar-Webster is professor of political science at State University of New York at Brockport and has served as codirector of the State University of New York–Paris Program. She is currently organizing a conference on cross-boundary pollution with Soviet, U.S., and Canadian participants. She received her Ph.D. from Columbia University. She is author of *Women under Communism, Czechoslovakia and the Absolute Monopoly of Power: Case Study of Political Power in a Communist Country*, "Ecology and Self-Management in Yugoslavia," and "Environmental Politics in Eastern Europe in the 1980s." She was a Guest Scholar at The Woodrow Wilson Center in 1987–88.

Zoltán Király specializes in research on plant pathology at the Plant

Protection Institute in Budapest. He is the chairman of the Plant Protection Committee of the Hungarian Academy of Sciences and editor of *Acta Phytopathologica*, a quarterly journal of mostly English-language contributions. Király, who received an honorary doctorate from the Agricultural University of Keszthely, currently teaches at the faculty of horticulture at the University of Budapest. He is author of *The Biochemistry and Physiology of Infectious Plant Disease*, *Methods in Plant Pathology*, and *Physiology of Disease Resistance in Plants*, as well as numerous articles.

John M. Kramer is professor of political science at Mary Washington University in Fredricksburg, Virginia. He received his Ph.D from the University of Virginia and has been a Research Fellow at the Russian Research Center at Harvard University and a Senior Fellow at the National Defense University. He is the author of *The Energy Gap in Eastern Europe* and a number of articles including "Nuclear Power in Eastern Europe" and "The Environmental Crisis in Eastern Europe: The Price for Progress."

Jerzy Kurbiel is on the staff of the Institute of Sanitary Engineering of the Cracow Technical University. Educated at the Technical University of Wrocław, the Warsaw Technical University (Ph.D. in 1966 in sanitary engineering), and the Robert Taft Sanitary Center in Cincinnati, Ohio, he specializes in water pollution control and wastewater treatment. He is author of some 130 scientific and technical articles and papers and coauthor of two textbooks. He has cooperated with the Environmental Protection Agency since 1973 on textile wastewater treatment and wastewater reclamation.

Winfried Lang, who holds a Ph.D. in international law and international relations from the University of Vienna, is the deputy head of the private office of the Austrian foreign minister. He has served as president of two United Nations conferences, one on the depletion of the ozone layer and the other on bacteriological and biological warfare. His publications include *International Environmental Protection, International Regionalism, Integration and Cooperation in North and South*, "Environmental Protection, Disarmament, and Law of Treaties," "Economic and Political Integration," and "Negotiations on the Environment."

Philip D. Lowe teaches at the Bartlett School of Architecture and Planning, University of London. He received his M.Phil. in history from the University of Sussex. Among his publications on countryside planning and its politics are "British Ecology in the 20th Century" and "The Withered Greening of British Politics: A Study of the Ecology Party." He is currently conducting research on Green politics in the United States. He was a Fellow of the The Woodrow Wilson Center in 1985.

Edward Mainland, a retired U.S. Foreign Service officer, is currently the chief of the Romanian Service of the Voice of America in Washington, D.C., specializing in environmental topics. He was previously the chief of the European Division of the Voice of America. He has written a number of articles on the environment in Eastern Europe, specifically the Danube delta, and contributed to *US-Romanian Relations, the First Sixty Years.*

Miklós Persányi is a senior advisor at the Hungarian Ministry of Environment and Water Management. Formerly the head of the Department of Environmental Information and Education at the National Authority for the Protection of the Environment in Budapest, he earned a degree in biology and public education from the Lorand Eötvös University of Sciences in Budapest and graduate degrees in environmental engineering and political science from the University of Budapest. He has written about environmental policy and environmental awareness and education in Hungary and Eastern Europe.

Helmut Schreiber is a research associate at the Institute for European Environmental Policy in Bonn, where he works on East and West policies and technologies related to the environment. He is a scientific advisor to the Heinrich Böll Foundation in Bonn. He is the author of *Umweltprobleme in Mittel- und Ost-Europa, Environmental Policies in Mature Industrial Regions: A Cross-National Comparison,* and *Urban Policy and Equality: The Case Study of West Berlin,* as well as numerous articles on environmental problems in Europe.

Charles E. Ziegler is associate professor of political science at the University of Louisville, specializing in Soviet domestic and foreign policy. He is currently an International Affairs Fellow at the Council of Foreign Relations in New York. He has written *Environmental Policy in the USSR* and is finishing a monograph on Soviet policy toward East Asia under Gorbachev for the International Institute of Strategic Studies in London. His "A Soviet Special Economic Zone" is a part of a Johns Hopkins Foreign Policy Institute project on U.S.-Soviet relations submitted to the president of the United States in a series of recommendations on U.S.-Soviet relations.

INDEX

accidents: chemical, 44–45, 198; explosions, 119; FRG/GDR agreement, 140; nuclear, 39, 128–31, 140, 141–42, 180. *See also* Chernobyl nuclear power plant

acid pollution, 3, 6; Bulgaria, 166–67; Hungary, 200–202, 207–8, 213; Romania, 235; transboundary, 125–26, 135–36

activism: antinuclear, 96, 107, 110; church, GDR, 9, 44, 96, 188, 190–92, 193; emergence, 4–12, 25–26, 43–49, 105–8; Greenpeace, 10; Hungary, 215–16; independent groups, 43–49, 50, 218–20, national, 111–14; underground press, 43, 44, 47; Western Europe, 103–8, 110, 111–14. *See also* dissidence

agency responsibilities, 28–31, 36; Bulgaria, 167, 168–69, 170, 171; FRG/GDR relations, 137, 138; Hungary, 198, 211–12; Poland, 225; Western Europe, 106–7, 109, 113–14, 121n.27; Yugoslavia, 29–30

agriculture: Bulgaria, 161–63, 172; Danube, 149, 150–51; fertilizers, 131, 182, 200–202, 203, 207–8; Hungary, 197–209, 212; GDR, 181–83; Poland, 229; Romania, 237; soil pollution, 3, 125, 199, 200–202, 207–8, 213. *See also* irrigation; pests and pesticides

air pollution: Bulgaria, 166–67; energy and, 3; GDR, 175–79; Hungary, 212, 213; Poland, 229; Romania, 235–36; Soviet Union, relations with, 89–91, 124; transboundary, 124, 125–27, 132n.2–33n.2, 135–36, 138–39, 175; Western Europe, 108, 114, 115–16. *See also* acid pollution

art and artists, 42, 84, 181, 188

Austria, 39, 46, 128–29, 218–19

automobiles, 6

Baltic Sea, 127–28, 223

Barcelona Agreement, 117

Basic Agreement on Environmental Cooperation, 138, 142

Berlin, 136–37

bilateral agreements: Austria-Hungary, 39, 128–29, 218–19; FRG-Czechoslovakia, 141; FRG-GDR, 39, 135–43; hydroelectric, 88–89; *vs* multilateral agreements, 124; nuclear accidents, 128–29; Romania-Soviet Union, monitoring, 89; water pollution, 127

biological control, 204–7

Black Sea, 167, 238, 240

Bolivia, 240

Bulgaria, 7, 9, 10, 18, 41–42; activism, 48; agriculture, 161–63, 172; energy, 71, 163–64, 172; water resources, 17–18, 159–73

Canada, 132n.2

Ceauşescu, Nicolae, 234

censorship, 27–28, 34, 40, 43, 46, 96; GDR, 175–76, 180, 183, 191; Romania, 235, 240

Charter 77, 42–43

Chernobyl nuclear power plant, 5, 8, 28, 43, 44; Czechoslovakia, response, 47–48, 70, 94; GDR, response, 40, 70, 96, 180–81, 192; Poland, response, 39–40, 70; Yugoslavia, response, 49

church activism. *See* religion

climate and weather, 159–60, 166

CMEA. *See* Council for Mutual Economic Assistance

coal, 3, 6, 16, 39, 57–58, 61–64, 75, 91, 92, 95; GDR, 176–79; Poland, 62, 63, 64, 93–94, 229; Romania, 62, 63, 75, 235, 236; transboundary implications, 131, 135, 138; Yugoslavia, 248–49, 254

coastal zones, 167

computer science, 88, 171, 172